T0299947

Rainwater Harvesting for Agriculture in the Dry Areas

Rainwater Harvesting for Agriculture in the Dry Areas

Theib Y. Oweis
International Center for Agricultural Research in the Dry Areas (ICARDA), Aleppo, Syria

Dieter Prinz
Karlsruhe Institute of Technology/ Karlsruhe University, Karlsruhe, Germany

Ahmed Y. Hachum
College of Engineering, University of Mosul, Mosul, Iraq

CRC Press is an imprint of the
Taylor & Francis Group, an **informa** business
A BALKEMA BOOK

CRC Press/Balkema is an imprint of the Taylor & Francis Group, an informa business

Typeset by V Publishing Solutions Pvt Ltd, Chennai, India
Printed and bound in Great Britain by TJ International Ltd, Padstow, Cornwall

Published by: CRC Press/Balkema
P.O. Box 447, 2300 AK Leiden, The Netherlands
e-mail: Pub.NL@taylorandfrancis.com
www.crcpress.com – www.taylorandfrancis.com

Library of Congress Cataloging-in-Publication Data
Applied for

ISBN: 978-0-415-62114-4 (Hbk)
ISBN: 978-0-203-10625-9 (eBook)

Contents

Preface

Water harvesting for agriculture is a term that means different things to different people. Misconceptions and unclear definitions are common - water harvesting is often confused with irrigation, soil and water conservation or other technologies; while engineers and scientists may use different terminologies. This book tries to clarify this confusion by developing a standard concept, and a framework that puts the concept, the components and the different types of water harvesting in a universal context. This will help design and implement systems to use scarce agricultural water more efficiently.

There is growing awareness of the need to harvest rainwater. Many efforts have been directed to collecting water from house roofs for domestic use. But water harvesting for agriculture is equally critical, particularly in dry areas. Water scarcity in many countries is approaching critical levels. As more water is needed for domestic use, industry and tourism, agricultural water is re-allocated to these sectors – while food demand and food insecurity continue to grow. In many dry areas, a large proportion of scarce rainwater is lost through evaporation. Water harvesting can reduce these losses, and alleviate water-related stress in agriculture.

Another factor is climate change, which will lead to higher rainfall intensities in many regions. This will increase runoff rates and associated soil erosion, reduce soil water storage, and increase moisture stress on crops and vegetation. As this book describes, water harvesting is a simple, low-cost method that can enable resource-poor farming communities to adapt to climate change.

Despite the increased attention, and solid evidence of the benefits of water harvesting, adoption by farmers is still low. The reasons (technical, socioeconomic, institutional and policy issues) are many, complex, and differ from one situation to another. We believe that understanding the reasons for slow adoption is an essential first step. The book discusses factors that help or hinder the dissemination of water harvesting technology, and suggests specific measures to accelerate adoption.

Most countries have indigenous water harvesting systems, some of which are still functioning after centuries of use. There are many types of indigenous systems, all based on the principle of collecting runoff water and channeling (concentrating) it for beneficial use. Indigenous systems use traditional materials and tools which were appropriate for earlier times, but these can be improved using modern materials and tools. We show in this book how indigenous knowledge is still valid but modern tools, materials and methods can help develop new practices that are more economic, durable and practical. The book emphasizes that new water harvesting systems must aim to complement – not replace – existing systems. Water harvesting interventions should form part of a comprehensive plan for managing water and land resources. They should be socially acceptable, technically sound, economically feasible, and environmentally sustainable.

Water harvesting cannot be practised without sufficient runoff. Runoff for water harvesting is encouraged and in some cases, when it is very low, it can be induced artificially. Runoff usually leads to soil erosion – but we seek to increase runoff in order to conserve soil and water resources! The book explains this apparent contradiction: how water harvesting, despite involving runoff, reduces soil erosion and land degradation.

One would expect that any water harvested for agricultural purposes must be diverted from other uses. This book provides the insights that this may be true in some cases, where it must be avoided; however in most cases we are simply recovering water that would otherwise be lost. For example, in arid environments, over 90% of precipitation returns to the atmosphere through evaporation. We show how and why this process occurs and how water harvesting can recover a large portion of this water and make it available for beneficial uses.

Water harvesting provides direct benefits to farmers, herders and investors. It also provides substantial indirect benefits, in the form of environmental health (controlling soil erosion and desertification, supporting ecosystems, reducing flood risk) and social benefits such as creating employment, reducing migration to cities, and better health for rural households. The indirect benefits are difficult to quantify and less apparent to farmers and investors than direct production benefits. This might make water harvesting less attractive as an investment priority. Should the public sector support farmers to invest in water harvesting, to encourage adoption? How can we account for benefits which extend to the larger public? This book tries to answer these questions, and guide policy makers making decisions on water harvesting investments.

The book is organized into 10 Chapters.

- Basic concept, definitions, history (Chapter 1).
- Hydrology aspects, which are critical in the design and implementation of water harvesting systems (Chapter 2).
- Water harvesting techniques, traditional and modern (Chapter 3).
- Methods to induce runoff (Chapter 4).
- Identifying potential sites for water harvesting, identifying areas suitable for specific techniques (Chapter 5).
- Planning and design of water harvesting systems: soil-water-plant-climate relationships, rainfall-runoff relationships, topography, engineering (Chapter 6).
- Storage of harvested water, recharging groundwater aquifers (Chapter 7).
- Operation and maintenance of water harvesting systems, importance of beneficiary participation.
- Factors that determine success: cultural and social factors, local priorities, participation, equity, land tenure, water rights, diseases, in addition to economic, technological and political sustainability aspects (Chapter 9).
- Water quality issues, environmental considerations (Chapter 10).

The authors of this book are scientists working with, or consulting for, the International Center for Agricultural Research in the Dry Areas (ICARDA), a non-profit organization with operations in over 50 countries. The information presented here is synthesized from research projects in different countries, implemented by ICARDA and/or partner organizations, notably Karlsruhe University, Germany (now the Karlsruhe Institute of Technology) and Mosul University in Iraq.

We hope this book will be useful to a range of readers: national policy makers, donor agencies, researchers, students interested in natural resources and the environment, and other stakeholders in agriculture and rural development.

Acknowledgements

The authors gratefully acknowledge Ms. Rima Dabbagh for her tireless effort in organizing and formatting the material for this book. We also appreciate the help provided by Mr. Ajay Varadachary for editorial support and Mr. George Chouha for photos and art works. Many other people – too many to list here – have contributed to this book directly or indirectly by providing ideas, comments and constructive reviews. We would like to thank them all. We also would like to thank the International Center for Agricultural Research in the Dry Areas (ICARDA) for constant encouragement and support. Finally, many of the practices, research methods and results described in this book were developed, tested or refined in partnership with teams from developing country institutions. We are deeply grateful to them for enhancing and helping to share the knowledge base on water harvesting practices.

About the authors

Theib Y. Oweis is one of the world's leading experts in water harvesting and management for agriculture, and Director of the Integrated Water and Land Management Program at the International Center for Agricultural Research in the Dry Areas (ICARDA). He has MSc and PhD degrees in Agriculture and Irrigation Engineering and has spent over 35 years researching methods to improve water productivity in agriculture, especially under conditions of water scarcity. He has made seminal contributions in water harvesting, supplemental irrigation, deficit irrigation, use-efficiency measurement, salinity management, and other areas. Dr. Oweis has authored over 200 scientific publications including papers in international refereed journals, books, book chapters and conference proceedings. He has also made major contributions in training programs for young scientists from developing countries.

Dieter Prinz has a PhD in tropical crop science from Goettingen University, Germany, after studies in horticulture and agriculture. His 43-year career has included teaching, research and technical consultancies in 17 countries, dealing with agricultural productivity, small-scale irrigation, water conservation, soil erosion control and water harvesting. He has led field research projects in a number of developing countries; conducted education and training programs in Asia, Africa and Europe; lectured at German universities; and served as adviser to a number of scientific foundations, UN bodies and other agencies. He was Full Professor in Rural Engineering and co-Director of the Institute of Water and River Basin Management at Karlsruhe University (now the Karlsruhe Institute of Technology), Germany, retiring in 2008. Dr. Prinz has authored more than 200 scientific papers, book chapters and contributions to conference proceedings.

Ahmed Y. Hachum is Professor of Farm Irrigation and Water Management at the College of Engineering, University of Mosul, Iraq, where he teaches graduate courses in irrigation systems, farm water management, drainage engineering, simulation and mathematical modeling, optimization and system analysis. Dr. Hachum has a PhD in Agricultural and Irrigation Engineering from Utah State University, USA. His research interests include irrigation systems, water harvesting, supplemental/deficit irrigation and other areas.

Dr. Hachum has been Editor-in-Chief of the Al-Rafidain Engineering Journal for several years, and has authored more than 70 technical publications including papers, book chapters, technical reports, and two textbooks on irrigation management. He has served as adviser to the Ministries of Irrigation and Agriculture in Iraq and as visiting scientist and consultant to ICARDA for many years.

Symbols

A	The catchment area
a	The area cultivated or cropped
a	Cropped area
b	The top width of bund
d	Height of the bund
D	The effective root zone depth
E	The storage efficiency of water in the effective root zone depth of the cropped area
E	The volume of earth/stone work per meter of bund length
ET	Evapotranspiration
m	The rank of the event
M_w	The mass of water in the soil sample
M_s	The mass of the soil sample after oven drying at 105°C
N:	The number of events
N_g	Number of days with daily rainfall equal to or greater than TR
P	Precipitation
P_e	Exceedance probability
P_c	Cumulative probability
P_m	The percentage mass
R	Mean or constant intensity rain
RT	Retention capacity
R(t)	Variable (time dependent) intensity rain
r	The runoff depth
T	Return period
t_s	Time at which the soil surface becomes saturated
t_r	Time at which runoff starts
t_e	Time at which runoff ends
U	The seasonal or yearly crop consumptive use
V	Volume of earthwork per hectare of the system land area
z	Slope of the bund sides or faces

Abbreviations

ASTER	Advanced Spaceborne Thermal Emission and Reflection Radiometer
AW	Available soil Water
BCR	Benefit-Cost Ratio
CCR	Catchment-to-Cropping-Area Ratio
CGIAR	Consultative Group on International Agricultural Research
CLG	Contour Laser Guiding
CWR	Crop Water Requirements
CWU	Consumptive Crop Water Use
DEM	Digital Elevation Model
DTM	Digital Terrain Model
DP	Deep Percolation
E	Evaporation
FC	Field Capacity
GCC	Global Climate Change
GIS	Geographic Information System
ICARDA	International Center for Agricultural Research in the Dry Areas
ICRAF	The World Agroforestry Centre
IfSAR	Interferometric synthetic aperture radar, a radar-based remote sensing technology
LIDAR	Light Detection And Ranging, an optical remote sensing technology
MSRC	Maximum Seasonal Runoff Coefficient
NRCS	The USA Natural Resources Conservation Services
PW	Permanent Wilting point
RC	Runoff Coefficient
SCS	Soil Conservation Service
SPOT	System Probatoire d'Observation de la Terre
SSC	Surface Storage Capacity
T	Transpiration
TAW	Total Available Water
TAC	Technical Advisory Committee
TR	Threshold Rain
WANA	West Asia and North Africa
WH	Water Harvesting
WHC	Water Holding Capacity

Chapter 1

Principles and practices of water harvesting

1.1 INTRODUCTION

In the arid and semi-arid regions, annual precipitation is generally low relative to potential evaporation and is poorly distributed over the crop-growing season. When rain does fall it is commonly in intense storms. In these regions, precipitation alone is generally not enough to support low-risk crop production.

The non-uniform distribution of precipitation in these areas usually results in frequent drought periods during the crop growing season. These cause severe moisture stress on growing crops and reduce yields; they may even cause crop failures. Most rainstorms are of high intensity. When the intensity is greater than the soil infiltration rate, runoff occurs. Runoff greatly reduces the amount of water that infiltrates into the soil and hence less water becomes available to the crop.

Figure 1.1 shows what happens to precipitation under semi-arid conditions:

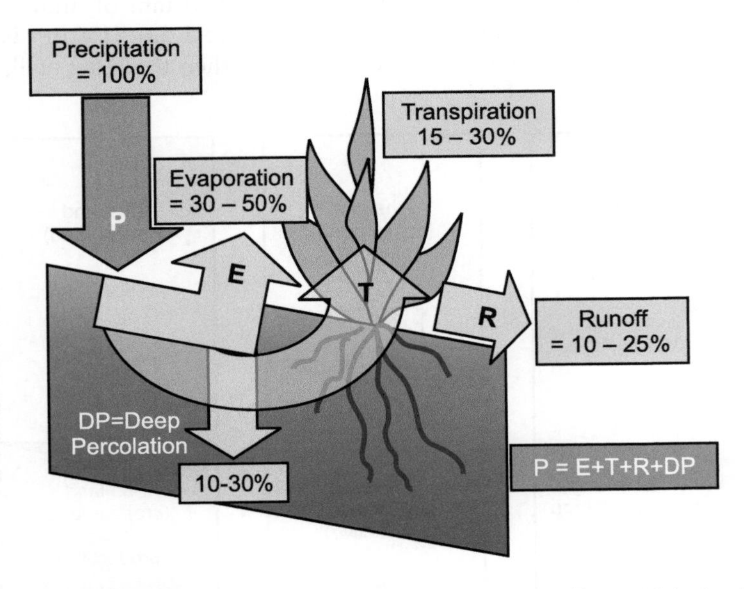

Figure 1.1 Average hydrological conditions in a semi-arid environment; P = precipitation, E = evaporation, T = transpiration, R = runoff, DP = deep percolation (Prinz & Malik 2002; based on Rockstroem & Falkenmark, 2000).

- A large part of the precipitation evaporates from the soil surface.
- Some of the water that infiltrates into the soil to a shallow depth also evaporates.
- Part runs off and is effectively lost to local crop production.
- Part is taken up by the roots of plants and returns to a large extent to the atmosphere via transpiration.
- The remainder replenishes the ground water.

The overall result is that most of the precipitation is lost back to atmosphere as evaporation or runs off, with no benefits to plant production. Other factors, such as degraded soils, steep slopes, poor vegetative cover, and unfavorable climate, aggravate the problem, reducing both water and land productivity.

Water harvesting is one option for increasing the availability of water to crops in dry areas. It increases the amount of water per unit cropping area, reduces the impact of drought and uses runoff beneficially (Barrow, 1999; Oweis *et al.*, 1999). Runoff irrigation, spate irrigation, and runoff farming are among the various forms and practices that come under the umbrella of water harvesting.

1.2 CONCEPT AND DEFINITION OF WATER HARVESTING

The basic principle of agricultural water harvesting is to capture precipitation falling on one part of the land and transfer it to another part, thereby increasing the amount of water available to the latter part. The objective is to provide enough water to crops on one part of the land to support economical agricultural production.

For example, land in an arid zone that receives 150 mm of annual precipitation cannot normally produce an economic crop (Figure 1.2a). If water is harvested on half of the land and transferred to the other half, then the latter will, in theory,

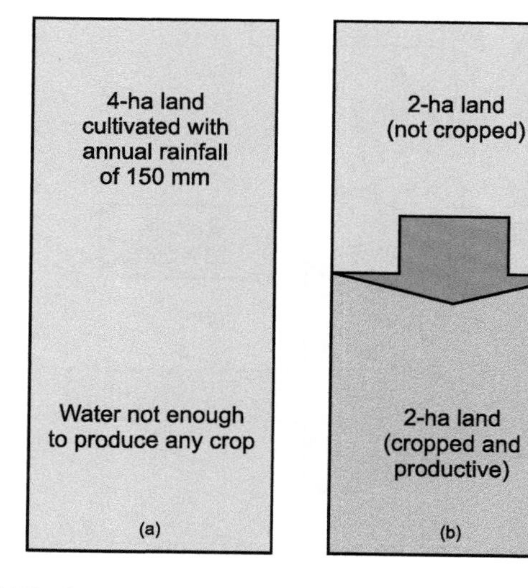

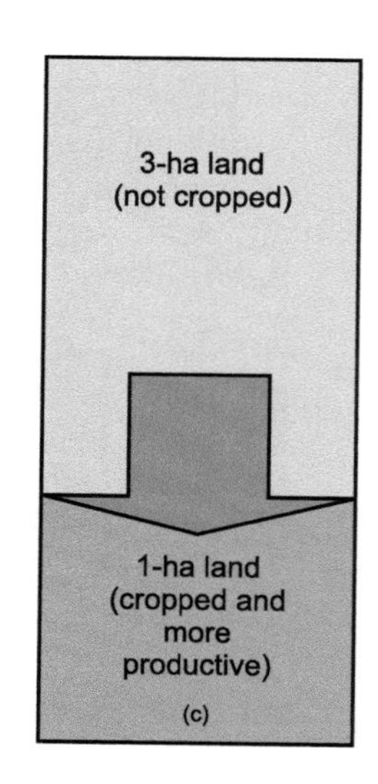

Figure 1.2 The basic concept of water harvesting for agriculture.

receive a total of 300 mm. This may be enough to support economic production of drought-resistant crops (Figure 1.2b).

If water were harvested on three-quarters of the land and this water were applied to the remaining quarter, this quarter of the land would receive a total of 600 mm of water, assuming a loss-free transfer of water (Figure 1.2c). This amount may be enough to support a wide range of crops that otherwise could not have been grown. Of course, it may be very costly to concentrate all of the precipitation from one part of the land to another, but some of this water may be diverted easily and cheaply. This will be discussed further in Chapter 4.

Such concentration of precipitation is called **water harvesting** (WH) and may be defined in various ways such as:

> "The process of collecting natural precipitation from catchments for beneficial use", or
>
> "Collecting and concentrating various forms of runoff from precipitation for various purposes", or
>
> "The process of concentrating precipitation through runoff and storing it for beneficial use".

Water harvesting is different from traditional soil & water conservation practices. Soil & water conservation practices aim at preventing surface runoff and keeping rainwater in place, whereas water harvesting makes use of, and even induces, surface runoff. With water harvesting, part of the land and most of the precipitation water will become productive. Agricultural production becomes possible and farmers' living standard will be improved.

Water harvesting may occur naturally or through human intervention. In dry areas natural water harvesting occurs when runoff from heavy rainstorms collects in depressions, which farmers then cultivate. Artificial water harvesting involves an intervention by farmers or others to improve precipitation collection and to direct runoff to cultivated land or reservoirs. More on the definition and principles of water harvesting may be found in Oweis *et al.* (2004a) and Nasri (2002).

Water harvesting systems in dry areas can provide water for domestic consumption, including drinking water; production of agricultural crops (including fodder bushes and tree crops); and livestock (Falkenmark *et al.*, 2001).

Water harvesting in dry areas also offers a number of environmental benefits (Figure 1.3), including reducing flooding risk, reducing soil erosion, reducing demand for surface water and groundwater, and recharging groundwater (Nilsson, 1988; Barrow, 1999; UNEP, 2009).

This publication focuses on meeting water demand of crops and livestock in dry areas.

1.3 HISTORY

The earliest water harvesting structures are believed to have been built 9000 years ago in the Edom Mountains in southern Jordan to supply drinking water for people and animals (Prinz, 1996). Simple forms of water harvesting for domestic and agricultural use are believed to have been practiced in the Ur area of Iraq over 6500 years ago. Runoff farming systems were used in the Negev desert over 4000 years ago. In this area, hillsides

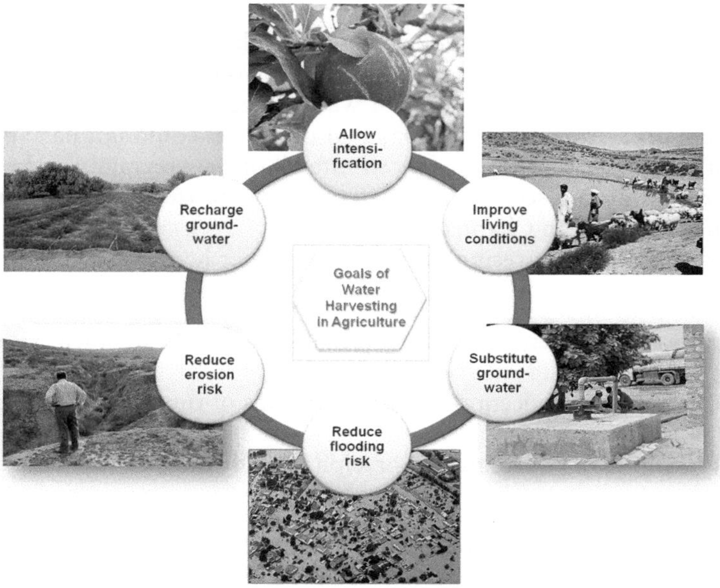

Figure 1.3 Goals of water harvesting in dry areas.

were cleared and smoothed to induce runoff, which was directed to agricultural fields. These systems played an important role in the successful establishment of settlements in the desert. In northern Libya (150–400 mm annual rainfall), runoff irrigation supported sustainable farming systems for over 400 years during the Roman Empire. *Sylaba* and *khushkaba* are ancient water harvesting systems that are still practiced in Baluchistan, a subtropical area in West Pakistan. The *meskat* and *jessour* systems are still used in southern Tunisia to support production of olives, date palms and figs (Oweis *et al.*, 1999). So-called *Lacs collinaires* have been used in Algeria since ancient times. Oweis *et al.* (2004b) provide an overview of indigenous water harvesting systems in this region.

Traditional water harvesting (*warping*) techniques have been in use in China for more than 3000 years (see Figure 3.4.2) and in India (*tanks and khadin*) since the 15th century. In Somalia, the *caag* system was used to harvest runoff water in sloping and flat lands (Figure 1.4). The ancient *hafairs* in Sudan are still in use for domestic and livestock purposes as well as for the production of pasture and other crops.

Rock and earth bunds and stone terraces have been used to harvest water in Niger and Burkina Faso. The *zay* (pitting) system, often used in combination with bunds, is an old practice in West Africa (Figures 1.5 and 3.2.1). From Chad, various traditional water harvesting systems have been reported (FAO, 2001).

In ancient times, rooftop water harvesting was practiced throughout the Mediterranean region. The harvested water was usually stored in underground cisterns. Some sophisticated examples of rainwater harvesting facilities were discovered in the ruins of the Palace of Knossos (1700 B.C.), the center of the Minoan culture on Crete (UNEP, 1983). The Yerebatan Saray in Istanbul, Turkey, is probably the largest cistern in the Mediterranean region, capable of storing about 80 000 m^3 of water

Figure 1.4 The *caag* system in Hiraan region, central Somalia (150–300 mm annual rainfall) (Critchley *et al.*, 1992).

Figure 1.5 The *zay* (pitting) water harvesting system in Burkina Faso. The system concentrates runoff water in pits, where plants are grown. Photo courtesy E. Dudeck, GTZ. (*See color plate, page 237*).

(Bamatraf, 1994). The water is said to have been collected from roofs and paved streets and a sophisticated system of filters ensured its cleanliness.

In Yemen, a water harvesting system known as *Seqayat*—underground tanks—was used for irrigation or drinking purposes. Nowadays, *Khazzan* are also used in Yemen. These are modern structures consisting of non-roofed, rectangular tanks with capacities that may exceed 80 000 m^3 (Oweis *et al.*, 2004a).

In southwest USA and northern Mexico, several traditional systems of water harvesting have been identified. The ancient systems, often used for domestic as well as for agricultural purposes, included, in many instances, collection ponds, cisterns, small masonry dams, and diversion canals. The importance of these systems declined as demand for water outstripped their capacities to deliver. They were also displaced by groundwater-lifting pumps and pressurized water-supply networks for small and large settlements (Barrow, 1987; Vivian, 1974).

In recent decades, there has been renewed interest in water harvesting, particularly in arid and semi-arid areas, as a result of shrinking water resources caused by increasing standards of living and higher population pressure in the dry regions of the globe. This renewed interest has also led to increases in the understanding, implementation, and management of water harvesting (Ben Mechlia *et al.*, 2009; Falkenmark *et al.*, 2001).

1.4 COMPONENTS OF WATER HARVESTING SYSTEMS

Regardless of the purpose or the type, all water harvesting systems have the following components (Figure 1.6):

A catchment area: This is the part of the land from which some of the precipitation is harvested. Therefore, it is also called the *runoff area*. The catchment may be as small as a few square meters or as large as several square kilometers. It may be agricultural, rocky, or marginal land, or a rooftop, a courtyard, or a paved road.

A storage facility: This is where harvested runoff water is held until it is used by crops, animals, or people. Water may be stored above ground, for example in jars, ponds or reservoirs; in the soil profile as soil moisture; or underground in cisterns or as groundwater in near-surface aquifers.

A target: This is the user of the harvested water. In agricultural production the target is the plant or the animal, whereas in domestic use it is the people and their needs.

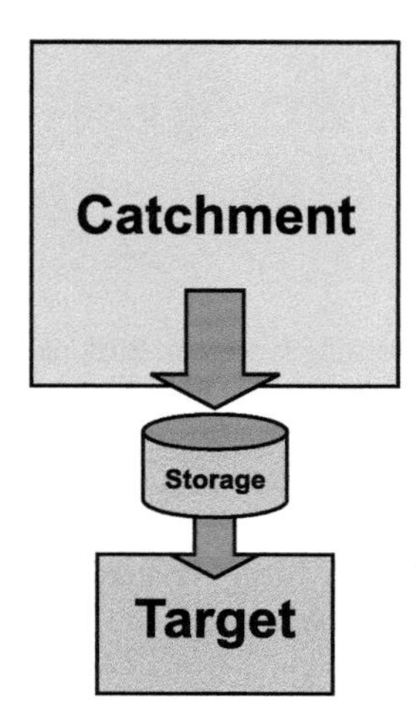

Figure 1.6 Major components of a typical water harvesting system.

Complex, large-scale water harvesting systems usually have an additional component for conveying and diverting runoff water to the target and/or storage facility.

Details of each of these components are presented in Chapter 3 (*Methods and techniques in water harvesting*) and Chapter 7 (*Storage of harvested water*).

1.5 IMPORTANCE AND BENEFITS OF WATER HARVESTING

Much of the rainwater that falls in arid and semi-arid regions is lost through evaporation or as uncontrolled runoff before it can be productively used. This is especially true in areas where all traditional water resources are already developed and used. Water harvesting in such areas could increase the water supply and help to reduce the problem of water scarcity. Improving the efficiency with which precipitation is used would also reduce pressure on traditional water resources and help alleviate the problems associated with present water shortages.

Water harvesting offers the following benefits (Oweis *et al.*, 2001; Prinz 2002b):

- In arid and semi-arid areas, water harvesting makes farming possible on part of the land, provided other production factors such as climate, soils, and crops are favorable. This is especially important when no other source of water is available for irrigation.
- Water harvesting systems can provide water to supplement rainfall to increase and stabilize production. It can alleviate the risk associated with the unpredictability of rainfall if the water harvested can be stored for later use in supplemental irrigation during drought periods. As the cropping risk is reduced, the application of organic or mineral fertilizer becomes economically viable, further increasing potential yields. Water harvesting can meet water needs for domestic uses and animal production where public supplies are not available.
- Water harvesting can provide water in arid areas suffering from desertification. This water can be used to increase vegetative cover and can help halt environmental degradation. It has also been found effective in recharging groundwater aquifers (Yahyaoui & Ouessar, 2000; Nasri, 2002; Somme *et al.*, 2004).
- Generally, water harvesting is a low-external-input technology and not difficult to implement. With a few exceptions, it does not require use of pumps or input of energy to convey or apply the water harvested.

Realization of these benefits leads to many intangible and indirect socioeconomic benefits such as: stabilization of rural communities; reduced rural-to-urban migration; utilization and improvement of local skills; and improvement of the standard of living of the millions of poor people living in drought-prone areas.

The implementation of water harvesting may, however, have a number of detrimental effects, such as:

- Increased soil erosion when slopes are cleared to promote runoff;
- Loss of habitat of flora and fauna due to clearance of slopes;
- Loss of habitat of flora and fauna in depressions (temporary wetlands);

- Conflicts among people living upstream and downstream of a watershed used for water harvesting;
- Conflicts between farmers and herders in dry environments (water in depressions is often used for livestock).

Dry-area ecosystems are generally fragile and have a limited capacity to adjust to sudden changes in the use of natural resources, especially land and water. The environmental consequences of the introduction of water harvesting may be far greater than can be foreseen. The introduction of water harvesting should, therefore, build on existing indigenous water conservation measures and should be implemented as one component of a larger village-level or regional (agricultural) water management improvement program (Malesu *et al.*, 2007; Ben Mechlia *et al.*, 2009; FAO, 1997).

1.6 IMPACT OF GLOBAL CLIMATE CHANGE AND ADAPTATION MEASURES

According to the United Nations Intergovernmental Panel on Climate Change, the average temperature of the Earth's surface rose by 0.7°C during the 20th century and is expected to increase by an average of about 3°C over the course of the 21st century, assuming greenhouse gas emissions continue to rise at current rates (Bates *et al.*, 2008). Even the minimum predicted temperature increase, 1.4°C, will represent a profound change that is unprecedented in the last 10 000 years.

The predicted changes in global climate include not only a rise in air temperature but also changes in rainfall regimes and a significant increase in the number of extreme weather events, such as stronger storms, longer droughts, and prolonged flooding (Figure 1.7 and Table 1.1). These phenomena aggravate the already existing

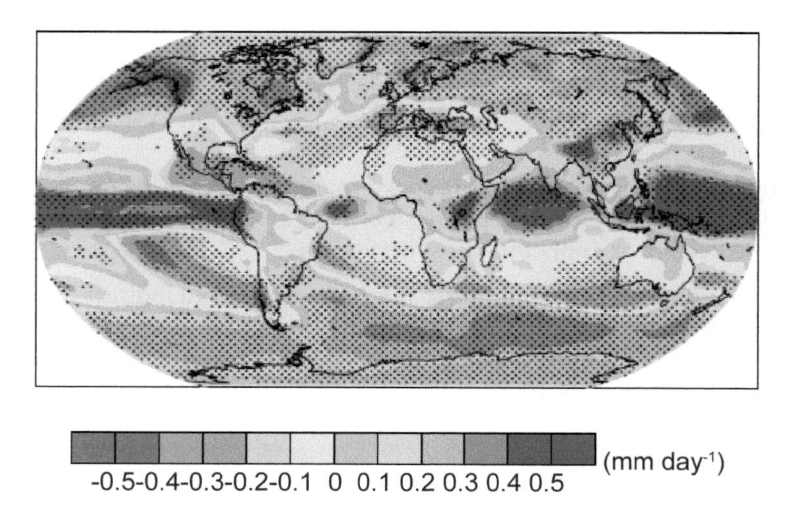

Figure 1.7 Predicted change in rainfall between the 20th and the 21st century (IPCC, 2007). (*See color plate, page 237*).

Table 1.1 Phenomena and impacts of global climate change (GCC), and adaptation measures relating to water harvesting.

GCC phenomenon	*Impact*	*Adaptation measures*
Increase in temperature	Shift of ecological belts by 20–58 km per decade	Shift of type of water harvesting techniques into zones where (minimum) average annual rainfall is given
	Higher water demand of crops and domestic animals	Increase catchment areas to collect more water
		Increase runoff coefficients on catchment areas
Change in rainfall regime	Shorter growing periods	Changes in crop selection/cropping pattern/ variety selection
	Rainfall more erratic	Increase (catchment area/cropping area) ratio
		Increase storage volumes
	Higher rain intensities	Strengthen/raise water harvesting structures (bunds, dams, walls)
Increased frequency of weather extremes	More/larger floods	Strengthen or increase size of structures for water diversion
		Strengthen and/or increase size of structures for impoundment and spill of excess harvested water
	More droughts	Increase (inter-annual) storage volumes
		Increase groundwater recharge (where feasible)
		Plant drought-tolerant crops and varieties
		Create alternative sources of income
	More storms	Plant more trees for protection of annual crops, domestic animals, and homesteads
		Provide stronger and more rigid WH structures

problems of farmers living in dry environments, such as erratic rainfall, frequent droughts, poverty, and inadequate resource availability (Pandey *et al.*, 2003; Bates *et al.*, 2008). Extreme rainfall events may also destroy water harvesting structures.

Climate change will require better management of rainwater, including water harvesting for domestic and agricultural purposes (ACPC, 2011; UNEP, 2009).

Chapter 2

Hydrological aspects of water harvesting

2.1 INTRODUCTION

The assessment of available water in a given area is the first step in planning, design, and operation of water-resource systems. Such systems may vary in size from micro-catchments to large reservoirs, depending on the need and available resources. The water-related activities may range from small-scale irrigation to watershed management for agricultural uses, including soil and water conservation and/or water harvesting. These variations require a variety of approaches and considerations in the estimation of water quantity.

The most important factors to be taken into consideration in planning water-harvesting interventions are climate, topography, soil and plant characteristics, hydrology, and socioeconomic conditions (Figure 2.1) (Tauer & Humborg, 1992; Prinz *et al.*, 1999). The climatic parameters such as temperature, rainfall, and evaporation greatly affect system hydrology (Strangeways, 2006). The likely future impacts of climate change on hydrological regimes also have to be taken into account (Van Dam, 2003; IPCC, 2007).

2.2 THE HYDROLOGICAL CYCLE

Water occurs in many places and many phases above, on and below the ground. The transformation from one phase to another and the movement from one location to

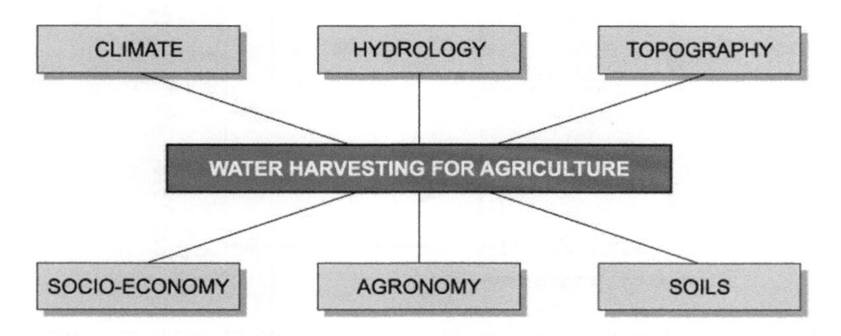

Figure 2.1 Basic factors to be considered in planning water harvesting interventions (Prinz *et al.*, 1999).

another constitute the hydrological cycle, which is a closed system. The total amount of water in the hydrosphere remains constant. However, global climate change will result in rainfall becoming more erratic and, in the southern hemisphere; more rain will fall over oceans and less over the land (IPCC, 2007).

Water balances can be drawn up for a region, or for an individual catchment area. This water balance will determine the optimal crop production from a given quantity of water.

Figure 2.2 shows the relationship between the various forms of water storage and water movement in a small catchment. The precipitable water in the atmosphere (W_i – W_o) is transformed and falls to the ground as precipitation (P). Some of the water on the surface of ground will infiltrate the soil through the surface of the soil (F), while part of it may find its way as overland flow (Q_0) into channel networks. Water may be transferred from the surface of the ground and from plant surfaces to the atmosphere by evaporation (E) or through vegetation by means of transpiration (T).

If rain intensity exceeds the infiltration rate or if the most upper parts of the soil matrix are saturated, rainwater will collect in puddles and then overflow on the soil surface. Surface runoff cannot occur, however, until a layer of water covers the path of motion. A portion of runoff may infiltrate into the ground or may evaporate, returning to the atmosphere.

During and following precipitation, soil moisture in the unsaturated subsurface zone is replenished by infiltration (F) through the surface. Once the upper layers are largely saturated, water will percolate to the deeper layers, recharging the groundwater (G). Some will also flow laterally through the soil (Q_i) – also known as interflow – into the channel network and contributes to stream flow during dry periods. During prolonged dry periods, soil moisture may be replenished through capillary rise (C) from shallow groundwater. Overland flow (Q_0), interflow (Q_i) and

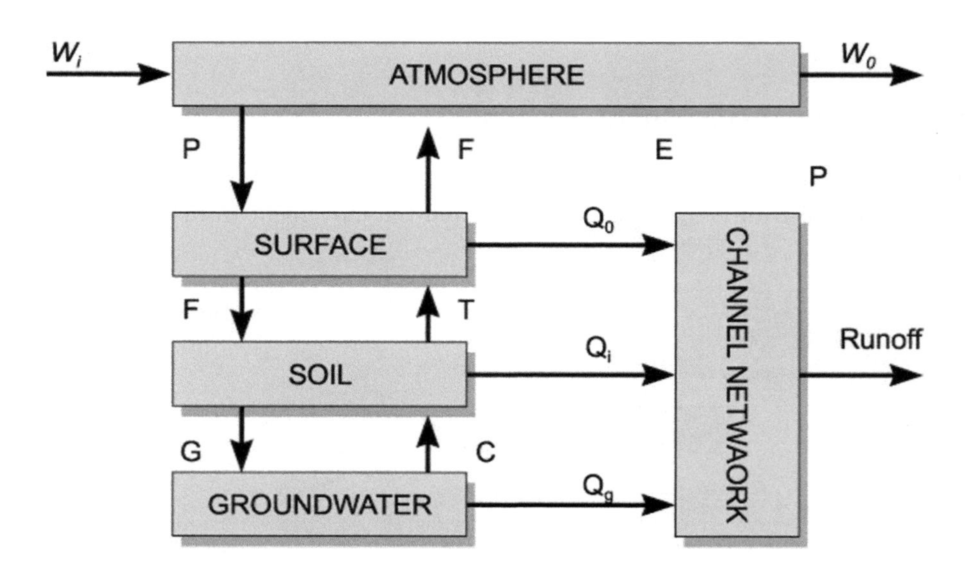

Figure 2.2 Relationship between the various forms of water and its water movement in a small catchment.

groundwater (Q_g) are all combined and modified in the channel network to form the runoff from the catchment.

In dry environments, most of the precipitation is in the form of rain. Therefore, the terms rainfall and precipitation are used interchangeably in this book.

2.3 SMALL HYDROLOGICAL WATERSHED MODEL

Hydrological modeling may be used to assess the amount of water available for agricultural uses in small watersheds in arid and semi-arid areas, where the rainfall is erratic in amount and in intensity.

The water that infiltrates into the ground first enters the soil zone that contains the roots of the plants. This water may return to the atmosphere through evaporation and transpiration. This upper zone can hold a limited quantity of water, the amount depending on the field capacity of the soil – the amount of water that a soil retains after drainage under gravity. If water is added to the zone when it is at field capacity, the water passes through to the groundwater zone. Water leaves the (near-surface) groundwater zone by capillary action into the root zone or by seepage into streams. The conceptual model for use at watershed catchment level can be represented as in Figure 2.3.

2.4 HYDROLOGICAL CHARACTERISTICS

The hydrological characteristics of a region are determined largely by its climate, topography, soil and geology. Key climatic factors are the amount, intensity, and frequency of rainfall, and the effects of temperature and humidity on evapotranspiration. Determining probable maximum precipitation, forecasting precipitation,

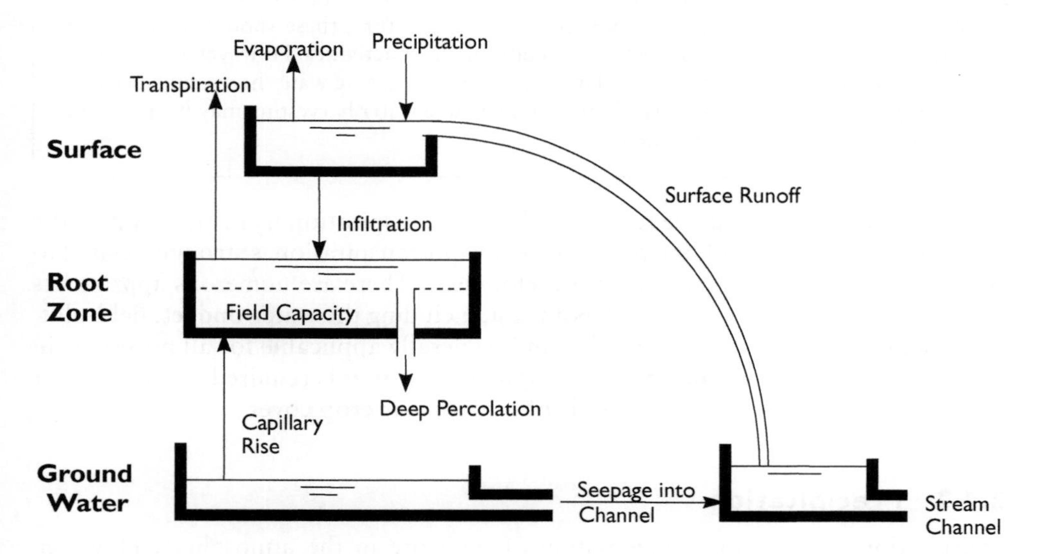

Figure 2.3 A hydrological model of a small catchment.

estimating evapotranspiration, and determining rainfall–runoff relationship can be problematic in arid and semi-arid areas.

2.4.1 Evapotranspiration

Evaporation and transpiration are decisive elements in the design of a water harvesting system for arid and semi-arid areas. Precipitation deposited on vegetation eventually evaporates, and quantity of water reaching the soil surface is correspondingly less than that observed in rain gauges. Evaporation and transpiration are indicative of changes in the moisture level of a basin. Estimates of these factors are also used in determining water supply requirements of proposed irrigation projects (Box 2.1).

Box 2.1 Crop water requirements and consumptive use.

The term 'Crop Water Requirements' (CWR) is commonly defined as the amount of water needed to meet the water lost through evapotranspiration from a disease-free crop, growing in large fields under non-restricting soil conditions, including adequate soil water and fertility, and achieving full production potential under the given growing environment.

This definition accommodates all processes impacting the water use by a crop, but excludes the influences of water stress, poor soil and fertility management, or inappropriate farming conditions. Thus, a complementary concept, 'Consumptive Crop Water Use' (CWU), is used when actual crop and field conditions are considered. CWU is defined as the amount of water utilized by a crop through evapotranspiration when grown under given farming conditions in a given growing environment. Therefore, when optimal cropping conditions are met CWU = CWR. Both CWR and CWU apply to both irrigated and rain-fed crops. However, CWR corresponds to potential yield production while CWU relates to actual crop growing conditions.

Typical CWR values for various crops grown on a variety of soils and under a range of environmental conditions are presented in the literature; these should be used as a guide for planning and implementing agricultural water activities. However, these values must be used with great caution in planning and implementing water harvesting, because the crop and field conditions usually prevailing where water harvesting may be necessary are commonly well below optimal.

Consumptive water use is the total actual evapotranspiration from an area plus the water used directly in building plant tissue. Evapotranspiration is strongly related to the density of cover and its stage of development. There are numerous approaches to estimate actual and potential evaporation, including the water-budget, field-plot, and lysimeter methods, but none of them is generally applicable for all purposes. In some hydrologic studies, mean basin evapotranspiration is required, while in other cases we are interested in water needs of a particular crop cover.

2.4.2 Precipitation

Precipitation results from condensation of moisture in the atmosphere. However, the general circulation, latitude, and distance to a moisture source are primary determinants of the climate in arid regions (Strangeways, 2006). The following are key terms used for rainfall analysis:

- ***Rainfall intensity*** is the quantity of rain falling in a given time over an area, and can be expressed in terms of cm/h or mm/h.
- ***Rain duration*** is the period of time during which a rainfall event takes place, and can be expressed in hours or minutes depending upon the duration and purpose.
- ***Frequency of rainfall*** is the frequency with which a given amount of rain falls over a given period, e.g. once in four years, once in six years, etc.
- ***Magnitude of rainfall*** is the total amount of rain falling at a point over a given period of time, i.e. daily, monthly, annually.

The rainfall characteristics in arid and semi-arid areas are different from those in temperate climates. In general, the rains are of high intensity, of shorter duration, and erratic in nature, with varying frequency and magnitude. The less the average annual rainfall, the greater, in general, is the variability. For designing a water harvesting system, the frequency of rain and probability of certain amounts and intensities are more important than the annual total. In planning water harvesting interventions it is desirable to have access to weekly or monthly rainfall records, but this is rarely possible.

For most water resources planning and design, including water harvesting studies, knowledge of precipitation over a definite area is required. The size of this area may vary from a small catchment of a few hectares to large river basins. As precipitation information collected with a rain gauge represents conditions at a point only, methods are required to transform point precipitation into entire area values (Linsley *et al.*, 1982). The three most common are the stations average method, the Thiessen polygon method, and the isohyetal method.

The stations average method is the simplest method, consisting of averaging all observed values. This method yields good estimates in flat country if the gauges are uniformly distributed and individual gauge catches do not vary widely from the mean. These limitations can be partially overcome if the topographic influences and areal representation are considered in the selection of gauge sites. This method assigns the same weights to each station regardless of location. This is a major drawback.

The Thiessen polygon method attempts to allow for non-uniform distribution of gauges by providing a weighting factor for each gauge. The stations are plotted on a map, and connecting lines are drawn. Perpendicular bisectors of these connecting lines form polygons around each station (Figure 2.4). The sides of each polygon are the boundaries of the effective area assumed for the station. The area of each polygon is expressed as percentage of the total area. Weighted average rainfall for the total area is computed by multiplying the precipitation at each station by its assigned percentage of the area and adding together the weighted averages for all the stations. The results are more accurate than those obtained by simple arithmetic method. Its major weakness is that it assumes linear variation of precipitation between stations and assigns each segment of area to the nearest station.

The isohyetal method is the most accurate method for averaging precipitation over an area. The locations of stations and rainfall recorded at them are plotted on suitable

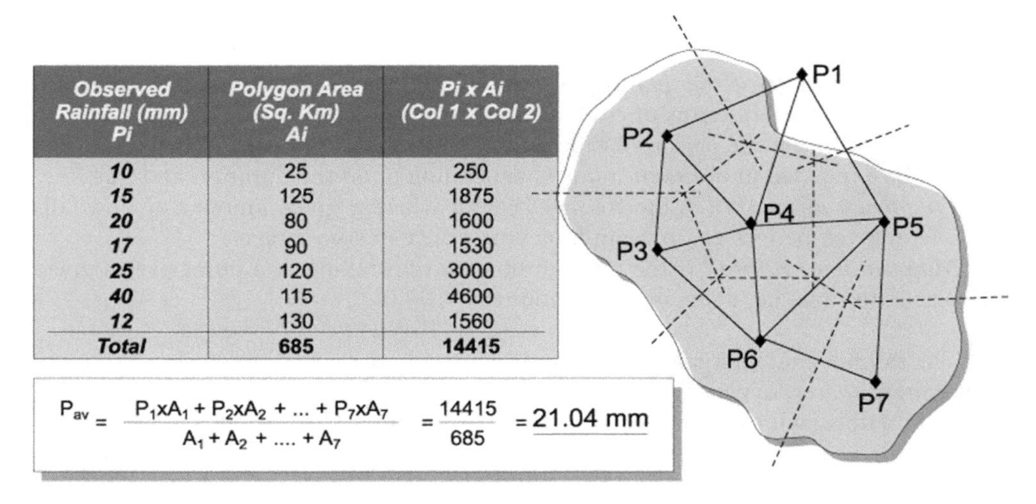

Observed Rainfall (mm) Pi	Polygon Area (Sq. Km) Ai	Pi x Ai (Col 1 x Col 2)
10	25	250
15	125	1875
20	80	1600
17	90	1530
25	120	3000
40	115	4600
12	130	1560
Total	685	14415

$$P_{av} = \frac{P_1xA_1 + P_2xA_2 + \ldots + P_7xA_7}{A_1 + A_2 + \ldots + A_7} = \frac{14415}{685} = 21.04 \text{ mm}$$

Figure 2.4 An example for application of Thiessen polygon method (Prinz *et al.*, 1999).

map and contours of equal rainfall are then drawn. Contours can be adjusted to take into account the effect of a mountain or a large hill or the direction of prevailing winds. The average precipitation for an area is computed by multiplying the average precipitation between successive isohyets by the area between the isohyets, summing these products and dividing by the total area.

Further details, demonstrations and application examples about these methods can be found in standard textbooks on hydrology (Linsley *et al.*, 1982; Chow *et al.*, 1988).

2.5 FREQUENCY ANALYSIS AND DESIGN RAINFALL

Frequency analysis can be used to estimate the frequency of occurrence of past events, or the probability of occurrence of future events. Rainfall is a continuous variable, varying with time, and can take any value greater than or equal to zero. Exceedance probability is the probability that the rainfall will be greater than or equal to a given value. For example, if the exceedance probability of 300 mm annual rainfall for a given location is 20%, one can expect that on an average the annual rainfall will be greater than 300 mm in 2 years out of 10, or one year out of five. This is equivalent to saying that in any year, the chance that the annual rainfall equals or exceeds 300 mm is one in five.

The return period or recurrence interval is the average time between occurrences of an event with a certain magnitude or greater. The return period T is related to exceedance probability P_e as follows:

$$P_e = 1/T \qquad (2.1)$$

Thus, for example, if the exceedance probability of a 250 mm annual rainfall for an area is 67%, the annual rainfall may equal or exceed 250 mm twice in a three-year period.

For water harvesting purposes, frequency analysis is usually performed for annual and monthly rainfall. Frequency analysis is made by plotting rainfall amounts against their cumulative probability P_c. The relation between P_e and P_c is:

$$P_e = 1 - P_c \tag{2.2}$$

For example, the P_e of zero annual rainfall in any location is 100% (i.e. $P_e = 1.0$), therefore, $P_c = 0$.

Plotting rainfall against P_e or P_c can be done in various ways. For water harvesting, it is sufficient to use the Weibull plotting position formula, because the required design value for rainfall lies within the range of the data. The Weibull formula is:

$$P_e = m/(N + 1) \tag{2.3}$$

where:

m is the rank of the event; m = 1 for the largest, and N for the smallest.
N is the number of events, such as annual rainfall.

For example, Table 2.1 presents the annual rainfall for 22 years, ranked in descending order. The third column in the table contains the exceedance probability according to Equation (2.3).

Table 2.1 Frequency analysis of annual rainfall at Khanasser valley, Syria, using the Weibull plotting position method.

Rainfall (mm)	*Rank (m)*	*P_e (%)*
399	1	4
387	2	9
335	3	13
315	4	17
293	5	22
291	6	26
249	7	30
244	8	35
238	9	39
235	10	43
223	11	48
213	12	52
194	13	57
182	14	61
174	15	65
155	16	70
154	17	74
150	18	78
109	19	83
106	20	87
98	21	91
93	22	96

The design rainfall is the amount of rainfall that is expected to be equaled or exceeded at a selected level of dependability (i.e. exceedance probability). In the example in Table 2.1, the design annual rainfall at a 70% exceedance probability (dependability) is 155 mm. This means that annual rainfall is expected to be 155 mm or more in 7 years out of 10. Similarly, design monthly or weekly rainfall can be evaluated.

Usually, 67% probability of exceedance is taken for the design of agricultural water harvesting systems.

2.6 RAINFALL–RUNOFF RELATIONSHIP

Surface runoff is generated when the rainfall intensity exceeds the infiltration capacity of soil for a period of time enough to get the soil surface saturated and puddled. Runoff is a function of many interrelated factors, such as soil type, soil moisture content, topography, land cover, and rainfall characteristics (intensity-time distribution).

2.6.1 Factors affecting runoff

Surface runoff is affected by many factors. The most important among them are soil type, rainfall characteristics, land cover, slope of the catchment area, and the size and shape of the catchment area.

2.6.1.1 Soil type

Coarse textured soils have stable structures and exhibit high infiltration rates, thus resulting in little or no runoff. Fine textured soils swell when wetted and shrink and crack upon drying. Infiltration rate is high initially, but falls rapidly to very low levels as the soil is wetted. Soils containing around 20% clay are highly prone to surface sealing, resulting in a crust or cap that makes the soil surface almost impervious. It is very important to take all these soil factors into consideration when planning and designing a water harvesting system.

2.6.1.2 Rainfall characteristics

Intensity, duration and frequency are the most important characteristics of rainfall for water harvesting. In dry areas, runoff-producing storms are usually of high intensity and short duration. The kinetic energy of falling drops is proportional to raindrop sizes and the total kinetic energy of a rainfall event increases with the increase of rainfall intensity. High-intensity rainfall breaks down soil aggregates at the soil surface, filling pores with fine particles. As a result, soil surface sealing develops which reduces infiltration and induces runoff. Therefore, runoff coefficients from intense, short-duration rainstorms are usually greater than those from less intense rainstorms having the same depth of rainfall.

2.6.1.3 Land cover

Surface roughness and vegetation impede surface water flow and increase surface storage capacity. Vegetative cover also protects the soil surface from the destructive effect of the falling raindrops. This reduces the development of crusting and soil surface sealing and hence reduces runoff.

2.6.1.4 Slope of the micro-catchment

Generally, runoff increases with increasing slope angle. This is mainly because less water is retained on the soil surface (SSC), and because the surface water flows more quickly towards the outlet, and hence less is lost by evaporation and infiltration in the catchment. Ground relief may take all kinds of shapes and sizes. For illustration purposes, Figure 2.5 shows the effect of slope on SSC assuming corrugations with a triangular cross section perpendicular to slope.

2.6.1.5 Size and shape of the micro-catchment

The runoff coefficient generally decreases with the increase in micro-catchment size and/or length of slope.

In a water harvesting system, rainfall induces surface flow on the runoff area, which preferably has a bare, crusted, and smooth surface. At the lower end of the slope, runoff water is collected in the target area. As rain starts to fall, part of the rainwater will be lost by infiltration in the runoff area, some where the water collects in shallow depressions, but also from soaking into the soil as it runs off. Besides runoff water, the target area also receives direct rainfall.

For macro-catchment, the rainfall-runoff process is extremely complex due to the large variation of natural conditions over the vast area of the runoff producing catchment (see Figure 2.7). Among these natural conditions/factors are the great variations, over the catchment, in soil type, topography, land cover, and land use.

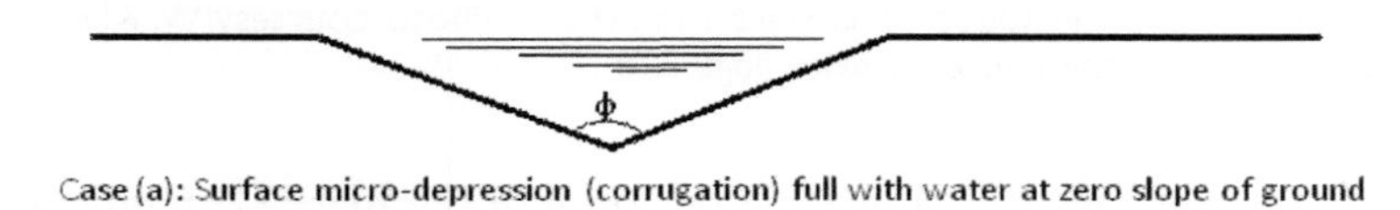

Case (a): Surface micro-depression (corrugation) full with water at zero slope of ground

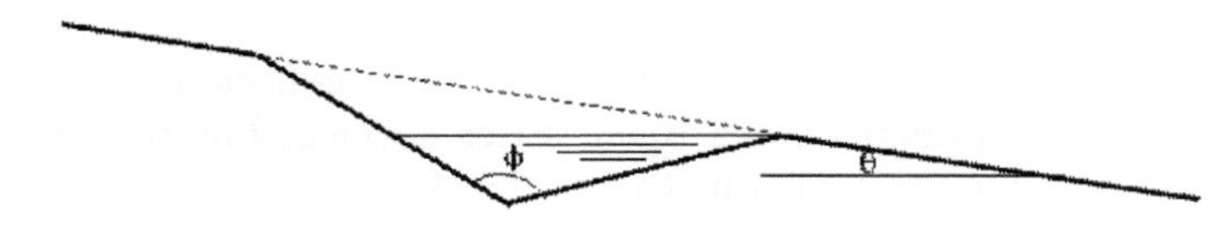

Case (b): Surface micro-depression (corrugation) partially full with water when ground slope=θ

Ratio of SSC of Case (b) to SSC of Case (a) = $\text{Cos}(\Phi/2+\theta)[\text{Sin}(\Phi/2+\theta) + \text{Cos}(\Phi/2+\theta)\tan(\Phi/2-\theta)]$; ($\theta \le \Phi/2$)

θ°	Slope (%)	Φ= 90°	Φ= 150°
3	5	0.9, (4.5 mm)*	0.34, (2 mm)
6	10	0.8, (4.0 mm)	0.22, (1 mm)

* Depth of SSC relative to 5 mm SSC for level ground surface (Case- a)

Figure 2.5 Effect of ground surface slope on Surface Storage Capacity (SSC) of rainwater.

Figure 2.6 Discharge measuring of macro-catchments is a precondition for hydrological modeling. Measuring station in Oued Mina region in Algeria. Photo courtesy W. Klemm/Karlsruhe University, Germany. (*See color plate, page 238*).

Therefore, it is important to monitor the response of the macro-catchment to major rainstorms by measuring the runoff at carefully selected gauging stations along the flow path of runoff (Figures 2.6 and 2.7).

While the rainfall is infiltrating and runoff is being collected in the target area, some water will evaporate from the open water surface, but the major portion infiltrates and is stored in the root zone (Boers, 1994).

2.6.2 Runoff models suitable for water harvesting

Rainfall–runoff models aim to describe surface runoff as a function of rainfall (Beven, 2000). The factors affecting surface runoff are soil type and moisture content, rainfall characteristics, topography, and soil surface cover and conditions. Model parameters are adjusted to a specific location and field conditions. Various methods are used to determine the relationship between rainfall and runoff (Tauer & Humborg, 1992), but most are suitable only for estimating surface runoff from large catchments. Here we focus on methods for estimating surface runoff from small catchments.

Figure 2.7 Catchment area for a macro-catchment project in Mali, West Africa. It is extremely difficult, to calculate the runoff volumes from slopes like this one, covered with various types of vegetation, rock outcrops, etc. Additionally, the water yield varies over the rainy season with changing vegetation cover. Photo courtesy W. Klemm/Karlsruhe University, Germany. (*See color plate, page 238*).

2.6.2.1 Runoff models for micro-catchment water harvesting

As rain falls, the soil surface water content gradually increases until it reaches a limit that depends on rainfall intensity. If rainfall intensity is less than the minimum infiltration capacity of the soil (the saturated hydraulic conductivity, P) or the rainfall duration is short, the soil surface layer will not become saturated and water will not pool on the soil surface. However, if rainfall intensity is higher, the soil surface will become saturated. Once the soil surface is saturated the rate of infiltration becomes less than the rainfall intensity and water will pool on the soil surface. When puddles and other depressions are filled (surface storage), surface runoff starts (Figure 2.8).

Figure 2.8 shows the response of a soil system under three different rainstorms. Storm 1 does not saturate the soil surface because the intensity is less than the saturated hydraulic conductivity of the soil, P. The second storm, also, does not produce runoff because the duration intensity that exceeded P was relatively short. Storm 3, however, saturates the soil surface by time t_s, satisfies the surface storage and generates surface runoff. Surface runoff has started at t_r and ended at t_e as indicated in the Figure. A curve for cumulative runoff is also shown. Between t_s and t_r, the surface storage capacity (SSC) has been filled.

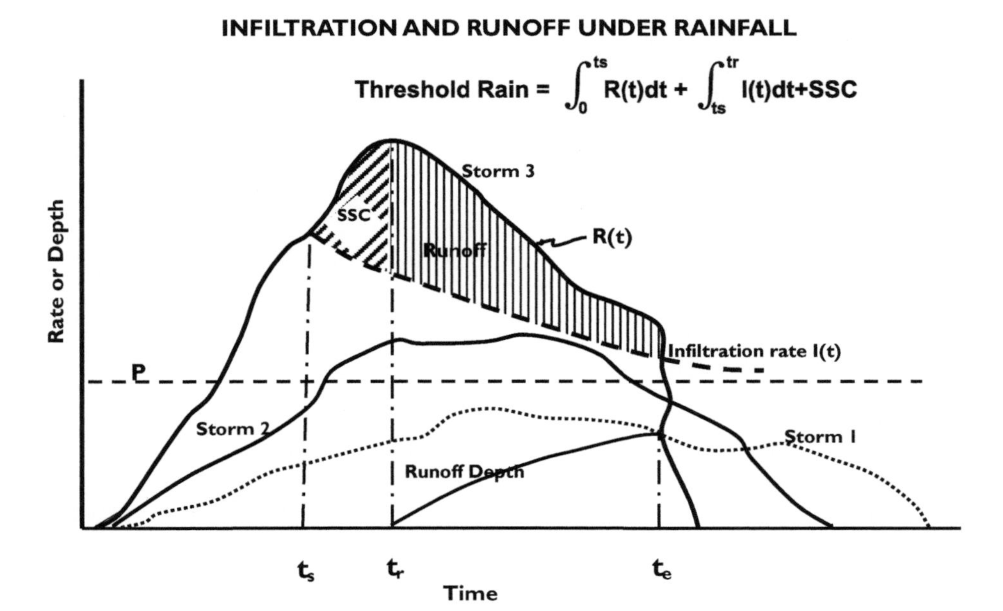

Figure 2.8 Micro-catchment response under three different rainstorms.

Threshold rain, TR, is defined as the total rainfall depth measured from the onset of the rainstorm until the start of surface runoff flow. TR is calculated using the following equation:

$$TR = \int_0^{t_s} R(t)dt + \int_{t_s}^{t_r} I(t)dt + SSC \tag{2.4}$$

where:

t_s is the time at which the soil surface becomes saturated
t_r is the time at which runoff starts
R(t) = rainfall intensity rate, function of time, t
I(t) = water infiltration rate, function time, t

Figure 2.9 shows one way to partition the rainwater on a micro-catchment into many components, based on a steady (i.e. constant intensity, R) rainstorm of duration T. Components 1, 2 and 3 are respectively the first, second and third terms in the right hand side of equation (2.4). Component 4 represents the amount of excess rainwater at any point in the micro-catchment. Component 5 represents infiltration, during rain, between t_r and T. Since the distance water travels in the micro-catchment of water harvesting is small and thus negligible, no allowance is made for routing surface runoff flow. Therefore, excess rainwater at all points in the micro-catchment is summed and taken as the generated runoff. Component 3 represents surface storage, which will

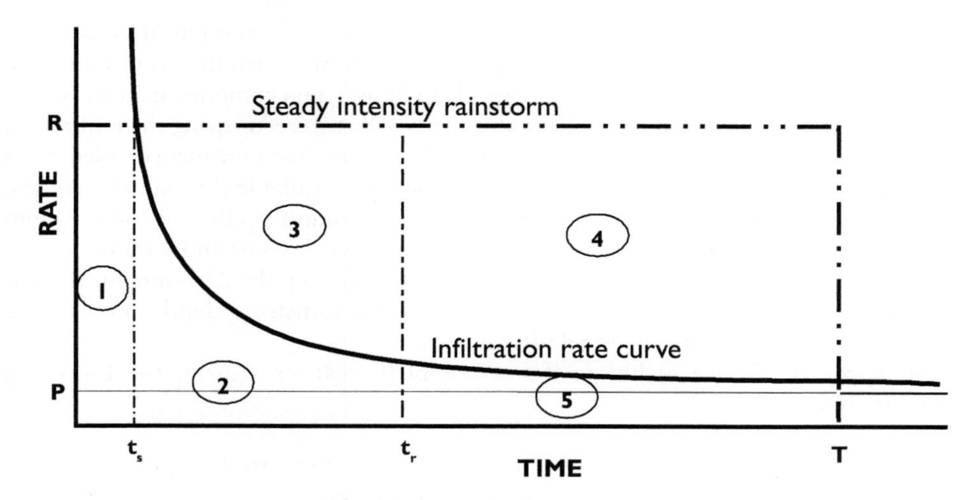

Figure 2.9 Components of partitioning a steady-intensity rainstorm of duration T for water harvesting purposes. Threshold rain is the sum of components 1, 2 and 3; Component 3 (above the curve) = surface storage; component 4 (above the curve) = surface runoff; Component 5 is the infiltration amount between t_r and T. P is the saturated hydraulic conductivity of the soil.

eventually infiltrate into the soil when the rain ceases or when the rain intensity becomes less than the infiltration capacity of the soil. Therefore, infiltration plays the major role in determining the amount of surface runoff. However, infiltration is a very complex phenomenon and extremely difficult to describe. It depends on numerous interacting physical, chemical, and biological factors, some of which are still unknown.

There are many models (empirical, physical, and numerical) to describe the behavior of the system (Hachum & Alfaro, 1980). However, factors affecting this system and its outcome, particularly prior soil moisture content, soil cover, and soil structure, are continuously changing during the growing season. Very few data on rainfall intensity and duration are available, especially in dry areas. Furthermore, it is extremely difficult, if not impossible, to predict the intensity and duration of future rainstorms. Therefore, the only rational way to estimate surface runoff for water-harvesting purposes is based on the depth of rainfall (daily, monthly, or yearly).

The runoff coefficient is defined as the ratio of the amount of runoff to the amount of rainfall. For a single storm, as shown in Figure 2.9, the runoff coefficient (RC) can be expressed as follows:

$$\text{RC} = \text{Component 4/rainstorm depth} \qquad (2.5)$$

The rainstorm depth can be expressed as:

$$\text{Rainstorm depth} = \text{TR} + \text{component 4} + \text{component 5} \qquad (2.6)$$

By combining equations (2.5) and (2.6) and rearranging them one gets:

$$\text{RC} = 1 - [(\text{TR} + \text{component 5})/\text{rainstorm depth}] \qquad (2.7)$$

By selecting a reasonable value for TR and assuming component 5 is very small (≈zero), one can get a maximum limiting value for RC in a given area. For example, if the depth of the rainstorm in Figure 2.9 is 10 mm and TR is taken as 4 mm, then the maximum value for RC is 0.60. For more rigorous, yet empirical, analysis, t_r and component 5 can be estimated to improve the evaluation of RC value. This will require knowing the infiltration function and assuming that the uniform rain rate equals depth of rainstorm divided by its duration. Procedures to estimate t_r and component 5 are available (Hachum & Alfaro, 1980). For illustrative purposes, Figure 2.10 shows the runoff coefficient for a 10 mm rainstorm falling at various uniform intensities and for two SSCs (0 and 2 mm).

Daily rainfall represents the sum of all rainstorms during the 24 hours of the day. However, for the purposes of the present analysis, the rainstorm depth in equations (2.5) and (2.7) will be taken as the daily rainfall.

To apply the above technique for a complete year or season, the following procedure is suggested:

1. Decide on a reasonable value for TR that is most suitable to the water harvesting site. This, of course, depends on land slope, cover, etc ...
2. Identify the number of days with daily rainfall equal to or greater than TR and denote it as N_g
3. Let T_1 and T_2 denote the total yearly rainfall and total rainfall for the N_g days, respectively.
4. Multiply N_g by TR and subtract the result from T_2 to get the sum of components 4 and 5 for the N_g days.
5. Calculate the maximum value of seasonal RC using the following equation:

$$\text{Maximum seasonal RC} = (T_2 - (N_g \times \text{TR}))/T_1 \qquad (2.8)$$

Tables 2.2 through 2.5 show the calculations of maximum seasonal RC (MSRC) for 22 years at Khanasser, Aleppo, Syria, for TR values of 2, 4, 6, and 8 mm. Since the

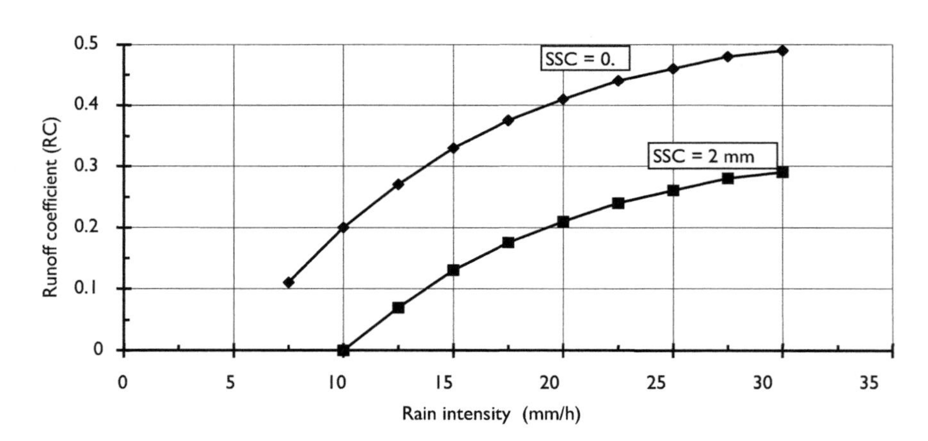

Figure 2.10 Runoff coefficient versus rain intensity for two storms having the same depth of 10 mm, assuming two surface storage capacity (SSC) values.

Table 2.2 Maximum Seasonal Runoff Coefficient (MSRC) for threshold rain of 2 mm at Khanasser valley, Aleppo, Syria, 1957/58–2000/01.

Season	*No. of rainy days*	*No. of days of runoff* (N_g)	T_1 *(mm)*	T_2 *(mm)*	*Maximum runoff (mm)*	*MSRC*[†]
1957/58	35	24	174	163	115	0.66
1959/60	27	13	98	84	58	0.59
1960/61	35	25	249	238	188	0.76
1961/62	47	31	235	217	155	0.66
1962/63	32	27	223	216	162	0.73
1963/64	33	31	182	180	118	0.64
1965/66	21	18	109	107	71	0.65
1966/67	51	49	387	384	286	0.74
1987/88	55	51	399	394	292	0.73
1988/89	18	18	150	150	114	0.76
1989/90	12	11	93	92	70	0.75
1990/91	24	24	244	244	196	0.80
1991/92	33	33	238	238	172	0.72
1992/93	22	22	213	213	169	0.79
1993/94	25	25	155	155	105	0.68
1994/95	30	28	194	192	136	0.70
1995/96	44	40	293	289	209	0.71
1996/97	46	43	315	312	226	0.72
1997/98	35	34	291	290	222	0.76
1998/99	17	17	106	106	72	0.68
1999–2000	29	27	154	152	98	0.64
2000/01	34	33	335	334	268	0.80
Average:	32	28	220	216	160	0.73

[†]Calculated using equation (2.8).

runoff coefficient ranges from zero to MSRC, one may take half of the average values given at the bottom of Tables 2.2 through 2.5 as recommended RC values for design purposes for the site under consideration. Design RCs for different TR values are presented in Table 2.6.

In small catchments, most of runoff is in the form of sheet flow, and hence runoff plots under controlled conditions may be used to measure runoff under rainfall of differing intensities. The plot must be representative for the area to be developed for water harvesting. It is advisable to experiment on plots of various sizes (slope lengths) and slope angles. Critchley and Siegert (1991) proposed a layout, materials, and procedure for testing runoff. At least two years of measurements are required to arrive at representative values for RC.

Overestimation of RC may result in reduced crop yields or crop failure due to water shortages (Rees *et al.*, 1991). Underestimation of RC results in setting aside more land than necessary as catchment areas and endangering the safety of the water harvesting system structures. The effect of excess moisture varies according to the crop. Millet, for example, can tolerate drought but not water logging but maize does not tolerate either.

Table 2.3 Maximum Seasonal Runoff Coefficient (MSRC) for threshold rain of 4 mm at Khanasser valley, Aleppo, Syria, 1957/58-2000/01.

Season	*No. of rainy days*	*No. of days of runoff* (N_g)	T_1 *(mm)*	T_2 *(mm)*	*Maximum runoff (mm)*	*MSRC*[†]
1957/58	35	14	174	133	77	0.44
1959/60	27	6	98	66	42	0.43
1960/61	35	20	249	225	145	0.58
1961/62	47	22	235	195	107	0.46
1962/63	32	22	223	203	115	0.52
1963/64	33	22	182	159	71	0.39
1965/66	21	12	109	91	43	0.39
1966/67	51	34	387	350	214	0.55
1987/88	55	37	399	359	211	0.53
1988/89	18	15	150	143	83	0.55
1989/90	12	6	93	81	57	0.61
1990/91	24	18	244	229	157	0.64
1991/92	33	25	238	216	116	0.49
1992/93	22	19	213	204	128	0.60
1993/94	25	18	155	137	65	0.42
1994/95	30	16	194	162	98	0.51
1995/96	44	27	293	257	149	0.51
1996/97	46	30	315	280	160	0.51
1997/98	35	24	291	265	169	0.58
1998/99	17	12	106	92	44	0.42
1999–2000	29	13	154	117	65	0.42
2000/01	34	27	335	317	209	0.62
Average:	32	20	220	195	114.8	0.51

[†]Calculated using equation (2.8).

2.6.2.2 *Runoff models for macro-catchment water harvesting*

Models suited to long-slopes water harvesting and floodwater harvesting include the unit hydrograph, the Soil Conservation Service (SCS) curve number, and the rainfall excess model. The first two methods are relatively simple, standard, well documented and can be found in most textbooks on hydrology and water-resources systems engineering. The third method is not as well known and is more complex.

A brief description of the first two methods is given below. More details and description of procedures for these methods may be found elsewhere (NRCS, 2008; Linsley *et al.*, 1982; Chow *et al.*, 1988).

Unit hydrograph method

The unit hydrograph method is still frequently used to determine runoff, despite many limiting factors. A hydrograph is a graph of discharge passing a particular point on a stream, plotted as a function of time. A unit hydrograph is a graph of the direct runoff

Table 2.4 Maximum Seasonal Runoff Coefficient (MSRC) for threshold rain of 6 mm at Khanasser valley, Aleppo, Syria, 1957/58–2000/01.

Season	*No. of rainy days*	*No. of days of runoff* (N_g)	T_1 *(mm)*	T_2 *(mm)*	*Maximum runoff (mm)*	*MSRC*†
1957/58	35	9	174	111	57	0.33
1959/60	27	3	98	52	34	0.35
1960/61	35	13	249	192	114	0.46
1961/62	47	17	235	172	70	0.30
1962/63	32	16	223	175	79	0.35
1963/64	33	10	182	102	42	0.23
1965/66	21	7	109	70	28	0.26
1966/67	51	23	387	297	159	0.41
1987/88	55	25	399	305	155	0.39
1988/89	18	8	150	111	63	0.42
1989/90	12	6	93	81	45	0.48
1990/91	24	14	244	212	128	0.52
1991/92	33	15	238	174	84	0.35
1992/93	22	15	213	186	96	0.45
1993/94	25	12	155	111	39	0.25
1994/95	30	11	194	141	75	0.39
1995/96	44	17	293	215	113	0.39
1996/97	46	19	315	231	117	0.37
1997/98	35	17	291	232	130	0.45
1998/99	17	8	106	74	26	0.25
1999–2000	29	10	154	105	45	0.29
2000/01	34	21	335	290	164	0.49
Average:	32	13	220	165	84.7	0.37

†Calculated using equation (2.8).

of 1 mm of effective rainfall distributed uniformly over the basin area (catchment) at a uniform rate during a storm of particular duration. The unit hydrograph is assumed to be representative of the runoff process for a watershed. The method is based on following three postulates:

- Constant duration of flow for a given drainage basin; the duration of flow depends on the duration of rainfall and not on its intensity.
- Linearity for rain of equal duration but of different intensity; runoff is proportional to the rainfall intensity.
- Superposition; runoff caused by several periods of rainfall can be superimposed.

The unit hydrograph should be derived from as many peak flows as possible. Monthly and annual mean or total flow is used to display the record of past runoff at a station.

One limitation of the unit hydrograph method is the assumption that storms occur with uniform intensity over the entire drainage basin. A unit hydrograph derived from

Table 2.5 Maximum Seasonal Runoff Coefficient (MSRC) for threshold rain of 8 mm at Khanasser valley, Aleppo, Syria, 1957/58–2000/01.

Season	*No. of rainy days*	*No. of days of runoff (N_g)*	*T_1 (mm)*	*T_2 (mm)*	*Maximum runoff (mm)*	*MSRC*[†]
1957/58	35	8	174	108	44	0.25
1959/60	27	2	98	46	30	0.31
1960/61	35	11	249	179	91	0.37
1961/62	47	13	235	146	42	0.18
1962/63	32	12	223	150	54	0.24
1963/64	33	5	182	70	30	0.16
1965/66	21	3	109	43	19	0.17
1966/67	51	18	387	267	123	0.32
1987/88	55	19	399	267	115	0.29
1988/89	18	7	150	105	49	0.33
1989/90	12	5	93	75	35	0.38
1990/91	24	9	244	179	107	0.44
1991/92	33	11	238	148	60	0.25
1992/93	22	9	213	146	74	0.35
1993/94	25	7	155	79	23	0.15
1994/95	30	9	194	127	55	0.28
1995/96	44	14	293	197	85	0.29
1996/97	46	11	315	177	89	0.28
1997/98	35	11	291	192	104	0.36
1998/99	17	6	106	62	14	0.13
1999–2000	29	7	154	86	30	0.19
2000/01	34	16	335	257	129	0.39
Average:	32	10	220	141	63.7	0.28

[†]Calculated using equation (2.8).

Table 2.6 Recommended seasonal runoff coefficient for planning micro-catchment water harvesting at Khanasser valley, Aleppo, Syria.

Threshold rain (mm)	*2*	*4*	*6*	*8*
Seasonal runoff coefficient	0.37	0.26	0.18	0.14

a single storm may not be representative, and it is, therefore, desirable to average unit hydrographs from several storms of about same duration.

Soil Conservation Service (SCS) method

The US Soil Conservation Service developed the curve number method to estimate the effect of land treatment and land use changes upon runoff (NRCS, 2008). It has been widely accepted and used as the method of choice for planning and design of soil and water conservation interventions. The popularity of this method is due to its simplicity, predictability, stability, and its responsiveness to watershed properties affecting runoff. The parameters used aim to quantify physical processes, although they may not be directly measurable. They usually represent spatially averaged catchment characteristics, such as surface cover type and conditions, soil type, and others. An important feature of the curve number method is that the proportion of rainfall converted into runoff (runoff efficiency) increases with the rainfall depth.

Box 2.2 Case study: Calculating retention capacity of traditional water harvesting systems in Tunisia.

People in the arid highlands of Tunisia have been harvesting water for centuries using a traditional system of dams known locally as *jessour* (singular = *jesr*).

Problem analysis: Low average annual rainfall (Medenine: 152 mm), high interannual rainfall variability (Medenine: 37–550 mm); a complicated hydrologic regime in watersheds, using the *jessour* technique for water harvesting.

Goals: Determine the water yield per *jesr* under various rainfall and topographic/pedological conditions.

Institutional framework: Researchers of the Institut des Regions Arides, Tunisia, in cooperation with a German researcher and the local agricultural extension service were working with farmers in this dry region to optimize water use.

Approach: In an area receiving an average of about 190 mm rainfall/year, hydrological (and economic) parameters were calculated based on interviews with farmers, aerial photos, hydrological studies, and application of a geographic information system (GIS).

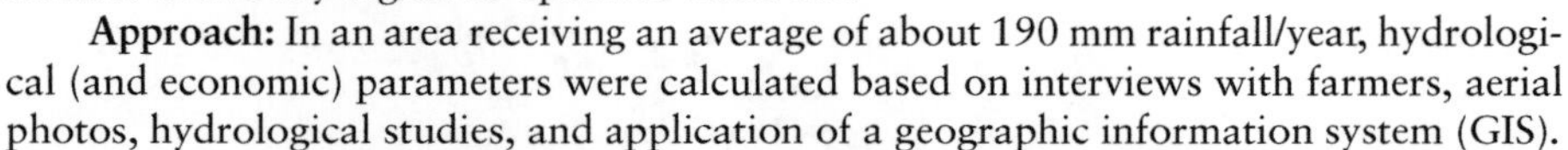

An aerial photograph (scale of 1:25 000) was used for spatial planning. Hydrologic units were identified, digitized, and processed using the Arc/View GIS software.

The watershed of Oued Saoudi was selected as the study area. The site, covering a total of 263 ha, consists of 105 *jessour* units (Figure B2.2.1).

The maximum retention capacity, RT_{max} (m^3), that can be stored behind the *tabia* (the dam of the *jesr*), is calculated as follows:

$$RT_{max} = H_{max} \times A/2$$

where:

H_{max} = maximum retention height (m);
A = retention surface (m^2)

Dividing the retention surface by two allows for the wedge-shaped form of the water volume; normally the terrace rises slightly upstream of the *tabia*.

Findings: The average retention capacity of the *jessour* was found to be 710 m^3, with a maximum of 7500 m^3.

Surplus water that cannot be stored behind a dam drains via a spillway into the downstream *jesr*.

Within the watershed, 75 000 m^3 of runoff can be stored on 105 *jessour*; these are primarily used for fruit-tree cropping. Retention of runoff also reduces floods and their associated damage (Figure B2.2.2).

Figure B2.2.1 Watershed of Oued Saoudi, southern Tunisia.

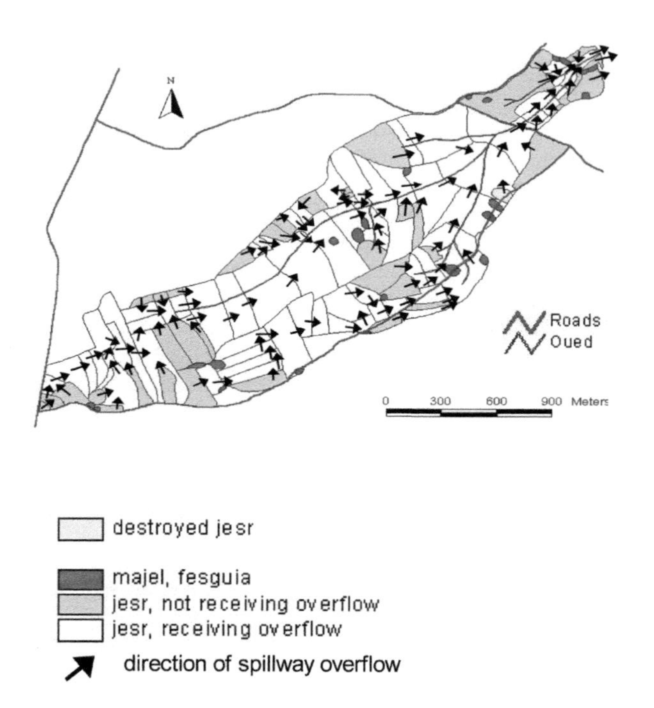

Figure B2.2.2 Types of water retention.
Source: Meinzinger *et al.* (2004).

Chapter 3

Methods and techniques in water harvesting

3.1 INTRODUCTION

Water harvesting is practiced in many ways to solve the various water needs of people living in the dry areas. Some of the techniques are used solely to provide water for plant production, while the others are used to provide water for human and animal consumption or for groundwater recharge (Figures 3.1 and 3.2). Some of the techniques in use are similar to each other, but may have different names in different regions. Others may have similar names, but be in practice completely different from each other (Critchley & Siegert, 1991).

Figure 3.1 Contour infiltration ditches combined with check dams (center), conserve soil and water and help recharge groundwater supplies. Photo courtesy A. K. Singh, Nirma University, India. (See *color plate, page 239*).

Figure 3.2 Channelling runoff water from long slopes (in the background) to fields, Kayes Province, Mali, West Africa. Photo courtesy W. Klemm/Karlsruhe University, Germany. (*See color plate, page 239*).

This chapter provides a general overview of the methods and techniques currently used to harvest water for agricultural and domestic purposes. These can be adopted with little modification in various dry regions of the world.

3.2 CLASSIFICATIONS OF WATER HARVESTING METHODS

There are several classifications of water harvesting methods, but the most commonly used system is based on the size of the catchment, i.e. micro-catchments and macro-catchments (Figure 3.3). Micro-catchment systems can be subclassified into rooftop systems and on-farm systems. Macro-catchment systems can be subclassified into long-slope systems and floodwater systems. The latter can in turn be subclassified into *wadi*-bed systems and off-*wadi* (diversion) systems.

The features of these various groups of techniques are given in Table 3.1.

Table 3.2 presents a classification proposed in this book for the various methods and techniques of water harvesting. The techniques described under the various

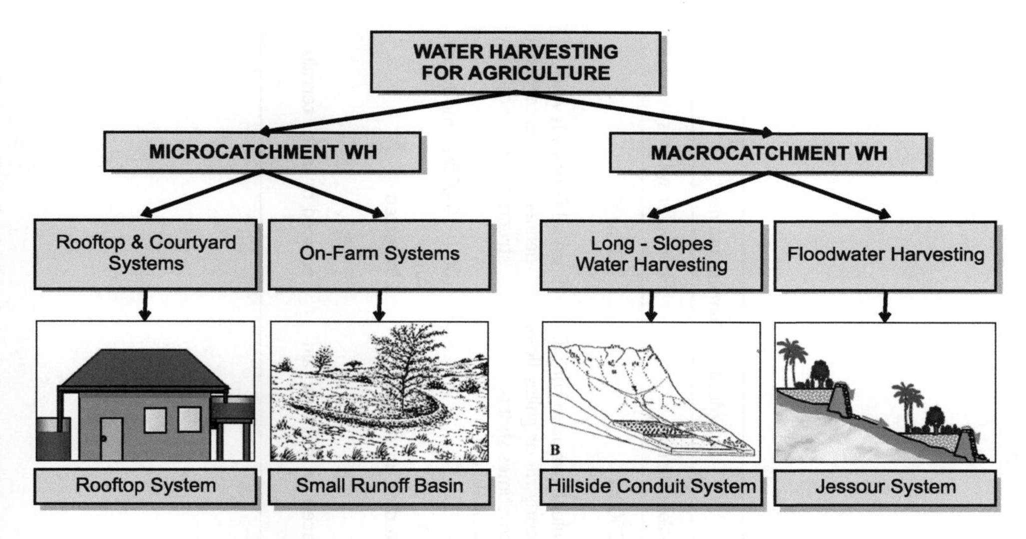

Figure 3.3 Overview and examples of water harvesting systems for agriculture. (Prinz, 2006; Rocheleau *et al.*, 1988; Prinz, 1996).

headings are not the only ones but represent the major techniques for various situations and productive uses.

3.3 MICRO-CATCHMENT WATER HARVESTING METHODS

Micro-catchment systems are those in which the surface runoff is collected from a small catchment area, where sheet flow prevails over short distances. Runoff water is usually applied to an adjacent agricultural area to be stored in the root zone and used directly by plants, or may be stored in a tank or cistern (in the case of roof top systems) for later use. Where the water is used for horticultural or agricultural purposes, the target area may be planted with trees, bushes, or annual crops (Figure 3.4). The size of the catchment may range from few square meters to over 1000 m^2.

The catchment surface may be natural, cleared or treated (with chemicals/compacted) to induce more runoff, or may consist of roofing material, plastic cover, or concrete.

Micro-catchment systems are simple in design and may be constructed at low cost. They are thus easily replicable and adaptable. They have higher runoff efficiency than macro-catchment systems and no water conveyance system is needed. Soil erosion is controlled and sediment directed to settle in the cultivated area. There are micro-catchment systems suitable to any slope and crop.

The most important advantage of micro-catchment systems is that the farmer has the control within his farm over both the catchment and the target areas, which is not usually possible in the case of macro-catchments. However, the catchment in

Table 3.1 Main features of the major groups of water harvesting methods and techniques.

	Micro-catchment systems		*Macro-catchment systems*	
Parameter	*Rooftop systems*	*On-farm systems*	*Long-slope systems*	*Floodwater systems*
Size of system (catchment + cropped area)	Area of available roof surface, <0.02 ha	<0.1 ha	0.1–200 ha	>200 ha
Predominant type of flow	Sheet or rill flow	Sheet or rill flow	Turbulent overland runoff, mostly rill or gulley flow, sometimes short-channel flow	Channel flow with well-defined course; complex structures needed
Catchment:cropping area ratio	Not applicable	1:1 to 25:1	10:1–100:1	100:1–10 000:1 (and more)
General slope of catchment area	0–50%	0–50%	5–60%	Any
Catchment surface	Galvanized corrugated iron sheets, corrugated plastic, tiles, cement surface, etc.	Usually treated	Treated or untreated	Untreated
Location of cropped area	Usually for household use	At lowest point of the system	Terraced or in flat terrain	Terraced or in flat terrain

Table 3.2 A classification of water harvesting methods and techniques.

	Micro-catchment systems		*Macro-catchment systems*		
				Floodwater harvesting	
Type of system	*Rooftop and courtyard systems*	*On-farm systems*	*Long-slope systems*	*Wadi*-bed systems	Off-*wadi* (diversion) systems
Technique	Treated surfaces (e.g. sealed, paved, compacted, smoothed surfaces)	Inter-row water harvesting, runoff strips Small runoff basins (*negarim/meskat*) Contour bench terraces Pitting techniques Contour bunds and ridges Semicircular, trapezoidal and rectangular bunds Vallerani-type water harvesting	Hillside conduit systems *Liman* Large semicircular-trapezoidal bunds Cultivated reservoirs/ tanks/*hafairs*	*Wadi*-bed cultivation *Jessour* Small farm reservoirs	Water spreading Tanks/*hafairs*
Type of storage	Cisterns, ponds, jars, tanks	Soil profile (ponds)	Soil profile, cisterns, ponds, reservoirs	Soil profile and ponds	Ponds, reservoirs, soil profile,
Aquifer recharge	None	Very limited	Limited	High	Very high

Source: After Prinz & Malik (2002).

Figure 3.4 Various micro-catchment water harvesting systems applied at a research station: Contour bund water harvesting (right); inter-row (runoff strips) water harvesting for grain, pulse, and forage crops (center); and semicircular bunds for forage bushes (left). The uncultivated areas serve as catchments. Photo courtesy T. Oweis/ICARDA. (*See color plate, page 240*).

this system occupies part of the farm area, and farmers will accept this only in drier environments. These systems generally require continuous maintenance and have relatively high labor requirements.

3.3.1 Rooftop and courtyard systems

Rainfall collected from rooftops is mainly used for drinking, especially in rural areas where tap water is seldom found (Worm & van Hattum, 2006). Between 80% and 85% of all measurable rain can be collected and stored. Indeed, by careful design, a family can live for a year from harvested rain in areas with as little rainfall as 200 mm per year (Morgan, 1990).

In general, water harvested from rooftops can be stored in tanks, jars, or underground cisterns (see also chapter 7). When collecting water from roofs and courtyards it might be advisable to install separate tanks for water intended for domestic use and that to be used on the garden (Figure 3.5). The runoff water collected may also be utilized to recharge groundwater if an infiltration well is built.

Development projects have implemented rooftop water harvesting in many tropical and subtropical countries such as Thailand and Ethiopia (Figures 3.6 and 3.7).

3.3.1.1 Suitable surfaces

Any roofing material is acceptable for collecting water. However, water to be used for drinking should not be collected from thatched roofs or areas covered with asphalt. Also lead should not be used in these systems. Galvanized, corrugated iron

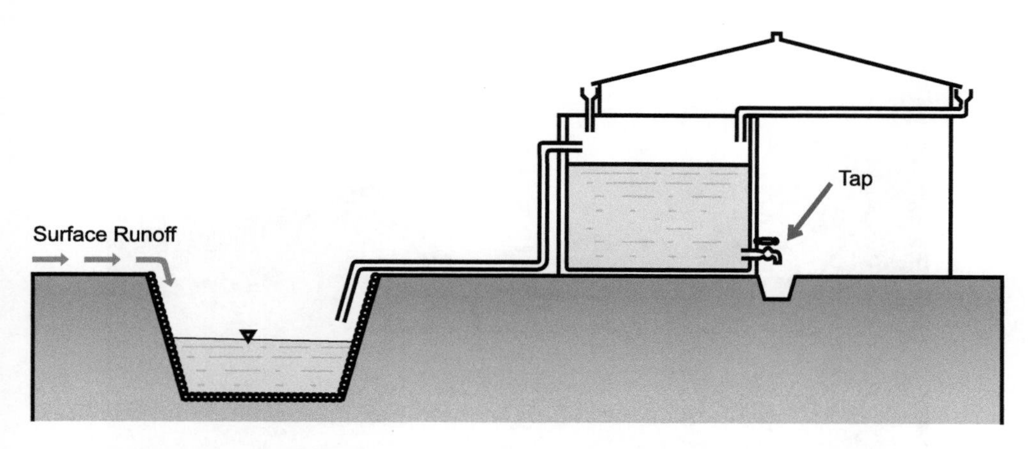

Figure 3.5 Two-tank system used to store water harvested from house roofs for domestic use and water from courtyard surface runoff for gardening purposes. (Pacey & Cullis, 1986).

Figure 3.6 In central Thailand, water harvested from rooftops is stored in ferro-cement tanks and used for drinking and to irrigate gardens. Photo courtesy D. Prinz/Karlsruhe University, Germany.

sheets, corrugated plastic and tiles make good roof catchment surfaces (Thomas & Martinson, 2007). Flat cement-covered roofs can also be used provided they are clean. Undamaged asbestos cement sheets do not have a negative effect on the water quality. Small damages may, however, cause health problems.

Roofs of greenhouses made of glass or plastic can be used to harvest water for irrigation (see Box 3, Figure B3.1.1). Concrete aprons, such as at airports, and roads can be used to harvest water for irrigation or for recharging groundwater supplies (Figures 3.8 and 3.9, Nissen-Petersen, 2006c.). However, roofs should be kept clean and water quality must be carefully monitored if it is to be used for drinking and domestic purposes. In modern times, roofs constructed with such materials like tiles or corrugated iron sheets are used in many regions of the world for harvesting water for domestic consumption (Water quality aspects are covered in Chapter 10.).

Figure 3.7 Rooftop rainwater harvesting in the Ethiopian Rift Valley. The Ethiopian Ministry of Agriculture supports construction of the tanks. Since 2006, all new school buildings and rural clinics in dry parts of Ethiopia must have rooftop water harvesting systems. Photo courtesy A. Hachum/ICARDA.

Figure 3.8 An example from Lanzhou, Gansu Province, PR China: A hilltop is covered with a concrete layer; the runoff from this artificial catchment is stored in 50 cisterns of 40 m^3 volume each. The water collected is used for irrigation of trees. Photo courtesy H. Hartung/FAKT, Stuttgart, Germany.

Figure 3.9 Use of roads as catchments in Ningxia Hui Autonomous Region, China. Photo courtesy D. Prinz/Karlsruhe University, Germany.

In industrialized nations, renewed interest in water harvesting has resulted in the equipping of modern buildings with water harvesting devices (for toilet flushing and irrigation) and in using greenhouse rooftops for water harvesting purposes. Although this technique is mainly used for domestic purposes, there are agricultural uses for it. One example is where plastic greenhouses are used to harvest the water. At a research project site at Egypt, the water collected from the roofs of greenhouses supplies about 50% of the water demand of crops in the greenhouse. Another example for this type of water harvesting technique is from Gansu Province, China (Box 3.1).

Box 3.1 Case study: Roof and courtyard water harvesting in Gansu Province, China.

Problem analysis: 250–450 mm annual precipitation; more than 70% of rainfall concentrated in three months; high population density; small landholdings; low income; no alternative water sources.

Goals: High water-use efficiency in economic terms; reliable income for farmers; high rainwater collection efficiency.

Framework conditions: The Provincial Government supplies technical advice, infrastructure, loans and subsidies; the farmers involved contribute labor.

Technical solutions: Rainwater from greenhouse roofs and from artificial catchments is collected in cisterns. Drip irrigation is used to grow flowers, vegetables (tomatoes, cucumbers, melons, etc.) and herbs in the greenhouses.

Achievements: The project has increased farmers' incomes and the supply of vegetables, flowers, and herbs to nearby cities.

Source: Yuanhong & Qiang (2001).

Figure B3.1.1 Only few greenhouses are covered with glass; most are covered with plastic. The runoff from greenhouse roofs is stored in cisterns located between the greenhouses. Electric pumps supply the water to the interior of greenhouses. Photo courtesy H. Hartung/FAKT, Stuttgart, Germany.

Figure B3.1.2 Drip irrigation is used to distribute the water within the greenhouse (Yuanhong & Qiang, 2001). (*See color plate, page 240*).

3.3.1.2 *Issues to be addressed*

The greatest limitation in the use of rooftops and other impervious surfaces for water harvesting is the cost of the design and construction (or purchase) of the storage facility. Options for reducing costs are discussed in detail in Chapter 7.

Much of the water that could be harvested may also be lost if the gutters are too small to handle the flow. The guttering system should be designed to cope with maximum anticipated water flow rates (Pacey & Cullis, 1986; Gould & Nissen-Petersen, 1999).

Care must be taken to avoid contaminating water that will be consumed by humans or animals. For example, the first flush of the new rains should be directed into a drainage channel and not captured in the storage structure. It may be necessary to pass the runoff through a sand filter to increase the water quality, especially where the quality of the harvested water is questionable. Air-borne insecticides and other pesticides and fertilizer may contaminate harvested water but the risk can be minimized by carefully choosing the location for the water harvesting structure. Storage structures should be kept as clean as possible and should be fenced to avoid access of domestic animals. They should also be covered to minimize evaporation and keep out animals such as mosquitoes and rodents (Heijnen & Pathak, 2007; Nissen-Petersen, 2007).

3.3.2 On-farm systems

There are numerous types of on-farm micro-catchment systems; the more important ones are described here.

3.3.2.1 Inter-row water harvesting

Inter-row water harvesting is used either on flat land or on gentle slopes of up to 4% having soil at least 1 m deep. It is suitable for areas with more than 200 mm average rainfall per year.

This system has two advantages:

1. It is the only water harvesting technique that can be used on absolutely flat land.
2. The construction can be fully mechanized.

On flat terrain (0–1% inclination) bunds or ridges are constructed and compacted using rollers or tractors. The soil can also be treated with sodium salts or other sealants to increase runoff (under higher-input conditions). The aridity of the location determines the catchment-to-cropping-area ratio (CCR). Generally, this ratio ranges from 1:1 to 5:1 (Figure 3.10).

Ridge height ranges from 30 to 100 cm and ridge spacing ranges from 1.2 to 10 m, depending on soil surface treatment, rainfall, and crop to be grown. Runoff is collected between the ridges and supplies a crop, but surplus water can be directed to a reservoir at the lower end of the system for storage and later use (see Figure 4.5).

Constructing the ridges may require large amounts of labor or machine input. Regular maintenance will also be required to maintain high rates of runoff. Depending on the type of soil, and the rainfall characteristics of the region, erosion control measures may also be needed.

The catchment area should be weeded and compacted regularly. The crop area should also be weeded to minimize competition for water. Crops commonly grown using this system include maize, beans, millet, grapes, and olives. Under this relatively high cost system, high value crops, such as fruit trees and vegetables, are more recommended.

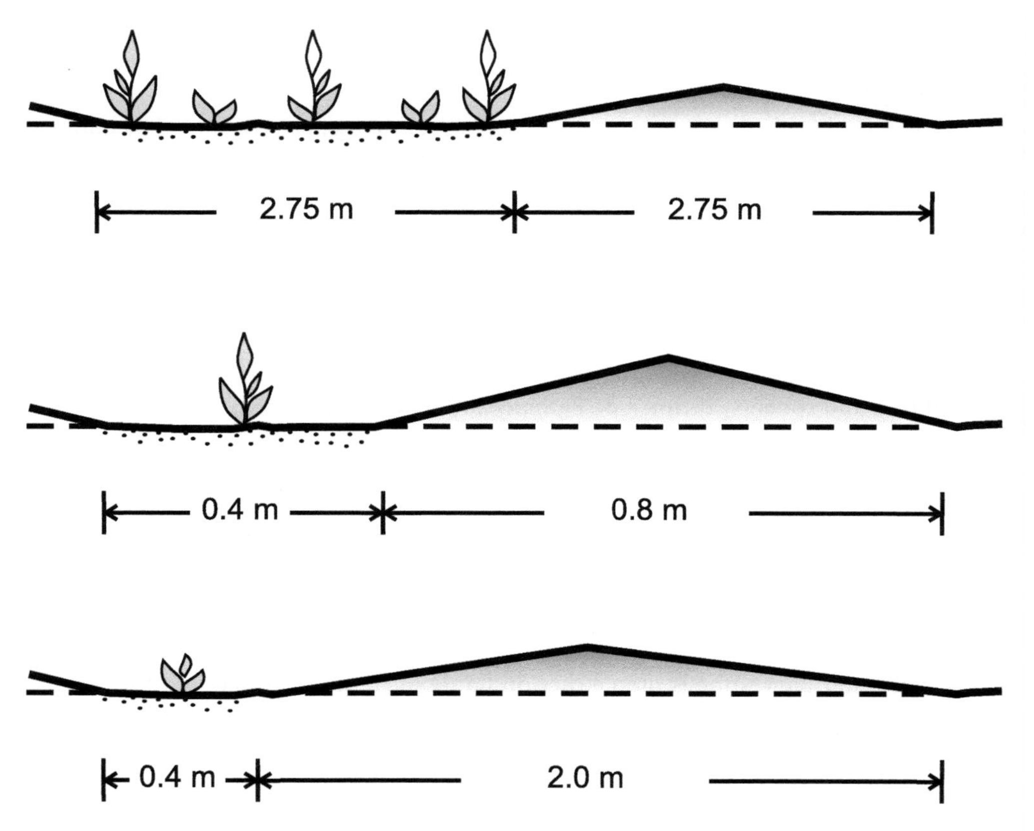

Figure 3.10 Inter-row water harvesting on flat terrain. The ratio of catchment to cropping area increases as rainfall declines. The ratio equals 1, 2, and 5 for the top, middle, and bottom case, respectively (Prinz, 1996).

On sloping land, the land is divided into strips along the contour; one strip is used as a catchment and the one downhill is cropped (Figure 3.11). The width of the cropped strip should be about 1–3 m, while the width of the catchment strip is usually of equal width, but may be up to five times as wide as the cropped strip, depending on the crop and rainfall. Runoff strip cropping can be fully mechanized and needs relatively little labor input. The cropped strips are cultivated every year. The catchment strip may be cleared and compacted to improve runoff.

In addition to the advantage of concentrating available water on the cropped strips, agricultural inputs are also concentrated on a fraction of the total land area. By continuously cultivating the cropped strip, soil fertility and structure improves and land become more productive.

However, one problem that farmers may face is that the distribution of water across the cropped strip may not be uniform. This happens especially on slight slopes when the cultivated strip is too wide, or if a small ridge is formed during cultivation along

Figure 3.11 Inter-row (runoff strips) water harvesting on sloping land. At the right hand side semi-circular bunds. Photo courtesy T. Oweis/ICARDA. (*See color plate, page 241*).

the upstream edge of the cultivated strip. To avoid this problem it is recommended that the cultivated strip should be no more than 2 m wide. Even water distribution may also be enhanced by forming small corrugations in the cropped area along the slope. These corrugations enhance the flow of surface runoff inside the cropped area (Figure 3.12).

3.3.2.2 Negarim

Negarims are small, diamond-shaped basins surrounded by low earth bunds (25 cm high). They are oriented such that the diagonal of the diamond is parallel to the slope of the land. Runoff flows to the lowest corner of the diamond, where an infiltration pit and a plant are located (Figure 3.13). The sides of the diamonds are commonly 5–20 m long, giving an area of 25–400 m^2. The size of the catchment and cropped area depends on the water requirement of the tree or bush species to be planted, land slope, soil type, and rainfall characteristics. Results have been poor with rainfall of less than 150 mm per year; hence this technique should be applied in regions with 150 to 500 mm rainfall. This technique and its variations are widely used in arid and semi-arid regions of the world, especially in sub-Saharan Africa.

Negarims can be constructed on almost any slope. However, soil erosion may occur on slopes greater than 10%, and bund height may be increased to an uneconomical

Figure 3.12 Runoff-strip micro-catchment water harvesting showing small corrugation in the cropped area used to increase the uniformity of water distribution. Photo courtesy T. Oweis/ ICARDA. (*See color plate, page 241*).

level. The CCR ranges from 1:1 to 25:1. *Negarims* are best suited to growing tree crops such as pistachio, apricots, olives, almonds, vine, and pomegranates, and to a lesser extent citrus fruits (Ben-Asher & Berliner, 1994). They are also used for rangeland rehabilitation, and for fodder bushes and trees for reforestation.

When used for growing trees, the soil should be deep enough (1.5–2.0 m) to allow adequate root development and to hold sufficient water for the whole dry season. If the catchment is well maintained, 30–80% of the rain could be harvested and used by the crop (Further technical details are given in Chapter 6). Soil conservation is a positive side effect of *negarims*. Once the *negarim* system is constructed, it lasts for years with little maintenance.

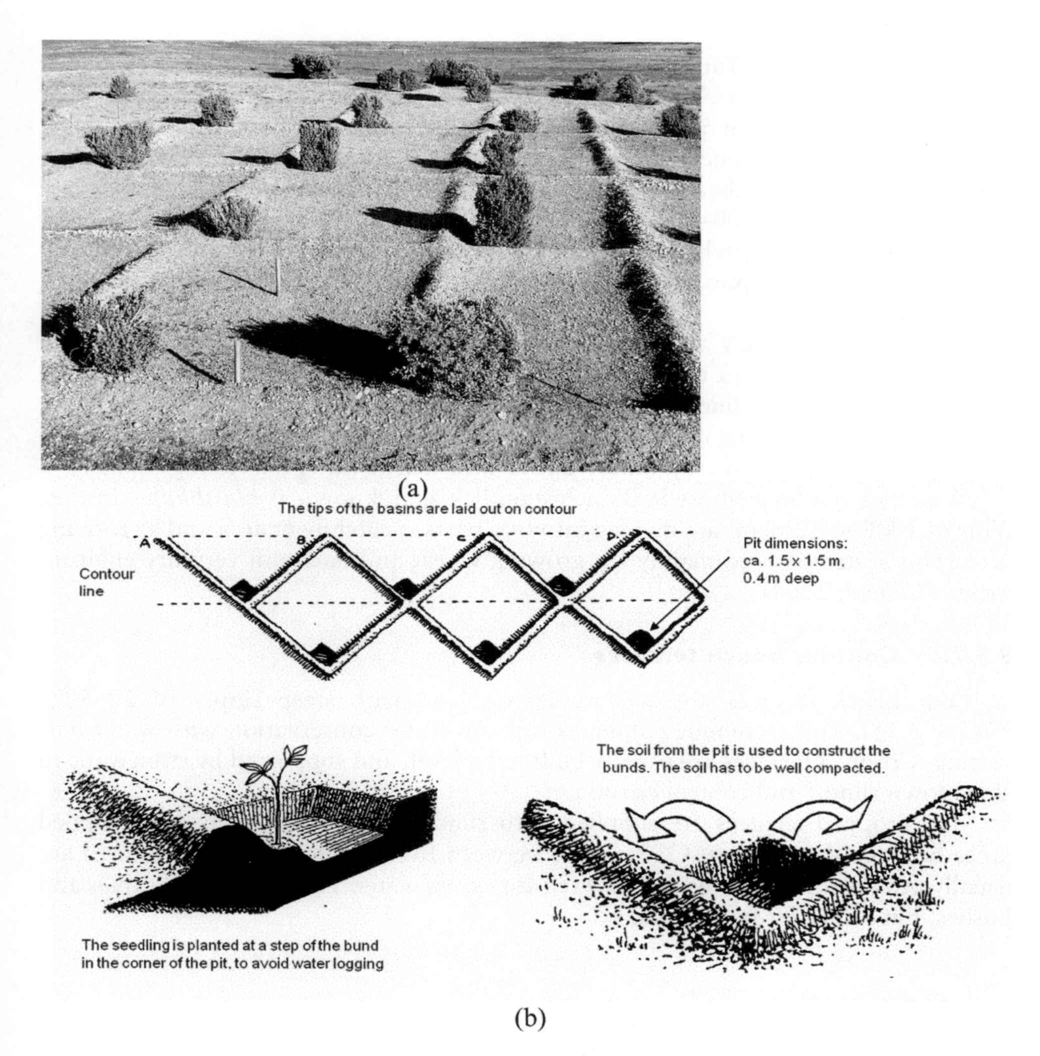
(a)
The tips of the basins are laid out on contour
A
B
C
D
Contour line
Pit dimensions: ca. 1.5 x 1.5 m, 0.4 m deep
The soil from the pit is used to construct the bunds. The soil has to be well compacted.
The seedling is planted at a step of the bund in the corner of the pit, to avoid water logging
(b)

Figure 3.13 *Negarim* micro-catchments: (a) Application in Syria; (b) Schematic drawing. Trees can be planted either at the bottom of the infiltration pit or on a step within the basin. (a) Photo courtesy T. Oweis/ICARDA; (b) Rocheleau *et al.* (1988).

Weed control needs to be done by hand. A high runoff coefficient has to be maintained. Since it supports high-value crops, applying runoff inducement treatments may be economically viable.

Evaporative losses may be high in the cropping area. Infiltration can be speeded up by loosening the soil in the cropping area and/or inserting an infiltration tube. Potentially productive areas are wasted because of the large distances between the trees. This problem can be overcome by planting trees in clusters and using about 1000 m^2 as the collecting area.

3.3.2.3 Meskat

Meskat is a term used in Tunisia, where this system is widely used (El Amami, 1983; Ben Mechlia & Ouessar, 2004). The *meskat* micro-catchment system consists of a catchment area (the *meskat*) of about 500 m^2 and a cropping area (*manka*) of about 250 m^2, giving a CCR of 2:1. The catchment and cropping areas are surrounded by a 15–30-cm-high bund equipped with spillways to let runoff flow into the cropping plots. *Meskats* are suitable for areas with 200–400 mm annual rainfall and land slopes of 2–15%.

A *meskat* has one catchment area but may have more than one cropped area, laid out in series so that surplus runoff water spills over from one cropped area to another one (Figure 3.14).

In North Africa they are used for growing olives, vines, figs, carob, dates, and barley. In Tunisia the area under *meskats* reached a peak of about 300 000 ha during the 1970s but has declined steadily since. Increasing population placed a higher demand on land, resulting in a decrease in the catchment area and a CCR of about 1:1. This increased the risk of crop failure due to insufficient water harvesting capacity.

A similar system is in use in Baluchistan, Pakistan, known as *khushkaba*. In this, plots of 1000 to 5000 m^2 are divided into two parts: a catchment area and a cropping area. This system is used mainly for growing wheat and barley in very dry environments (Ahmad, 2004).

3.3.2.4 Contour bench terraces

Contour bench terraces are constructed on land with steep slopes of 20–50% (Figure 3.15). This technique combines soil and water conservation with water harvesting. Cropped terraces are usually built to be level, and supported by stonewalls to slow down runoff and control erosion.

The cropped terraces are supplied with runoff water from steep, non-cropped areas between the terraces. CCR ranges between from 1:1 to 10:1. The terraces are usually provided with drains to safely release excess water. They are used for trees and bushes, but rarely for field crops.

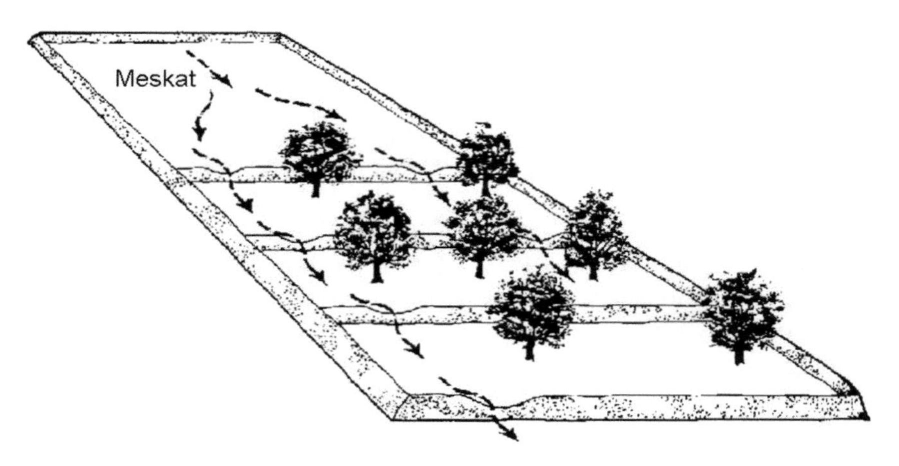

Figure 3.14 The *meskat*-type water harvesting. Adapted from (El Amami, 1983).

Figure 3.15 Contour bench terraces: (a) Example from southern Tunisia; (b) System on 65% inclination; (c) System on 14% inclination. Photo (a) courtesy D. Prinz/Karlsruhe University, Germany.

Terraces may be used in areas where the annual rainfall is between 200 and 600 mm. The construction can be done manually or using heavy machines. Construction costs and labor requirements for maintenance are high. If catchment surfaces are left bare, erosion may become a problem.

On milder slopes of 10% or less with deeper soils, conservation bench terraces might be the best technique among the contour terrace types (Figure 3.16). Unlike contour bench terraces, there are no supporting stone walls constructed due to lower land slope.

3.3.2.5 Small pits

The pitting system is a very old technique mainly applied in western and eastern Africa, but also adopted in some areas of West Asia and North Africa (WANA). The most famous pitting system is the *zay* system in Burkina Faso. This form of pitting consists of digging pits that are 5–15 cm deep and 0.3–2.0 m in diameter (Wright, 1985). Manure and grasses are mixed with some of the soil and put into the *zay*, improving the fertility and structure of the soil in the pits. The rest of the soil is used to form a small dike on the down-slope side of the pit. The *zay* system is also used in combination with bunds to slow runoff (Figure 3.17). Pitting systems are used mainly for the cultivation of annual crops, especially cereals such as millet, maize, and sorghum. Pitting is applied in areas with an annual rainfall of 350–600 mm. It is

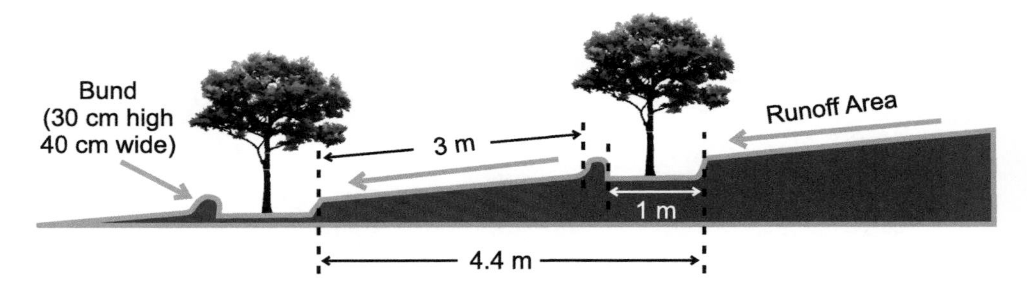

Figure 3.16 Conservation bench terraces constructed on 10% slope and planted to tree crops.

Figure 3.17 The *zay* pitting system in Burkina Faso, West Africa. The field is bunded to retain the runoff. Photo courtesy D. Prinz/Karlsruhe University, Germany. (*See color plate, page 242*).

commonly used on flat land or on slopes of up to 5%. The use of pits on flat land is regarded more as an *in situ* moisture conservation technique than a water harvesting system. Each pit has a catchment area of around 0.30 m^2 and a cropped area of about 0.1 m^2. The spacing between pits is 50 cm along the slope and 1 m across the slope.

Other pitting systems include the Kitui and Katumani systems in Kenya, *covas* in Cape Verde Islands and the Wamatengo system in Tanzania. Some of these systems may be also regarded as *in situ* water conservation techniques.

Labor requirement for pitting is high as the pits have to be reformed after each harvest.

3.3.2.6 Contour bunds and ridges

Contour bunds or ridges are constructed along the contour at spacing ranging from 5 to 20 m (Figures 3.18 and 3.19). One to two meters of the space between the

Figure 3.18 Contour ridges. Photo courtesy T. Oweis/ICARDA. (*See color plate, page 242*).

Figure 3.19 Contour ridge water harvesting with fodder bushes. Photo courtesy T. Oweis/ICARDA. (*See color plate, page 243*).

bunds or ridges is cultivated, while the rest is the catchment. The height of the ridges varies according to the slope and the expected runoff depth behind them. The bunds are made of packed soil, but may be reinforced with stones when needed. It is a simple technique, which can be carried out by the farmers themselves. Bunds can be formed manually, with an animal driven-implement, or using tractors with suitable implements. Ridges may be constructed on slopes from 1% to 50%.

The key to the success of these systems is to locate the ridge as precisely as possible along the contour. If this is not done properly, water will flow along the ridge, accumulating at the lowest point, eventually breaking the bund and destroying the whole system. A transparent tube 10–20 meters long, attached to two poles, and filled with water is a simple and efficient tool for identifying the contours; one person at either end moves the poles until they are at the same level. More details about this simple survey technique can be found elsewhere (Oweis *et al.,* 2001; Critchley & Siegert, 1991).

If precise contouring is not feasible, small cross-bunds (ties) may be put in every 2–5 m along the ridge to prevent flow of water along the ridge (Figure 3.20). The minimum length of a cross-bund is 2 m if the contour bund spacing is 5 m or more.

Contour ridges are ideal for supporting forage, grasses, and hardy trees in drier environments, and for sorghum, millet, cowpea, and beans in semi-arid tropics.

If planting trees, an infiltration pit should be excavated at the junction of the contour bund with the cross-bund. The recommended infiltration pit size is 80 cm by 80 cm, with a depth of 40 cm. It should be dug at least 30 cm away from the cross-bund. The earth excavated from the pit can augment the building of the contour bunds and cross-bunds. Contour bunds set 10 m apart with 2.5-m-long cross-bunds every 5 m are commonly used for tree plantations.

A special form of contour ridges may be constructed on mild slopes with stone bunds every 20–30 m down the slope (Figure 3.21). These stone bunds are permeable and serve only to slow sheet flow and allow for more infiltration. However, earth may

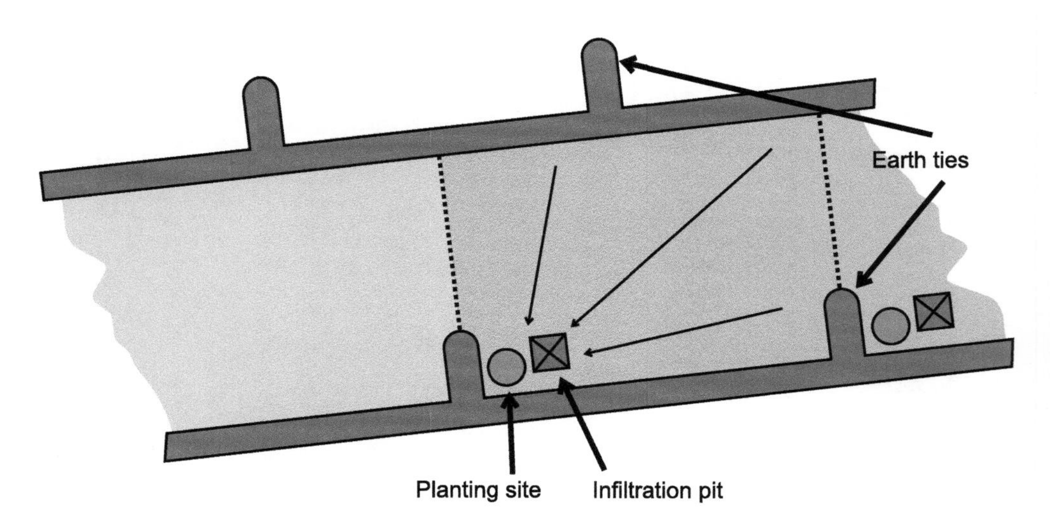

Figure 3.20 Contour ridges with cross-bunds. Adapted from (Critchley & Siegert, 1991).

Figure 3.21 Stone bunds used with *zay* water harvesting in Burkina Faso, West Africa. Photo courtesy D. Prinz/Karlsruhe University, Germany.

be excavated and added to the upstream side to form an impermeable contour ridge. In the semi-arid tropics this system is sometimes combined with the techniques like the *zay* system or *in situ* water-conservation techniques such as the tied-ridge system. These systems can only be used if stone is readily available near where it is needed.

Due to internal erosion between the bunds, soil will accumulate at the lower side of the strips. This can be reduced by sowing grass along the bunds prior to the rains. Once these grasses are established, they will have to be controlled to avoid them spreading to the catchment area.

To avoid erosion, storm water drains and/or controlled overflow by spillways may be necessary. Again, care should be taken during construction to ensure accuracy of the surveyed portion and adequate size of bunds in case of large volumes of water. If there is a risk of system damage from external runoff, a diversion ditch should be built along the upper border of the system to intercept surface runoff and direct it downhill.

3.3.2.7 *Semicircular and trapezoidal bunds*

Semicircular and trapezoidal bunds are set with the tips of the arms of the bunds set on the contour line, facing up the slope (Figure 3.22). The bunds are usually made of earth. They are created at spacings that allow sufficient catchment area to provide enough runoff water for the plant within the bund. These bunds are usually located in staggered rows as shown in Figure 3.22. The distance between the ends of the bunds

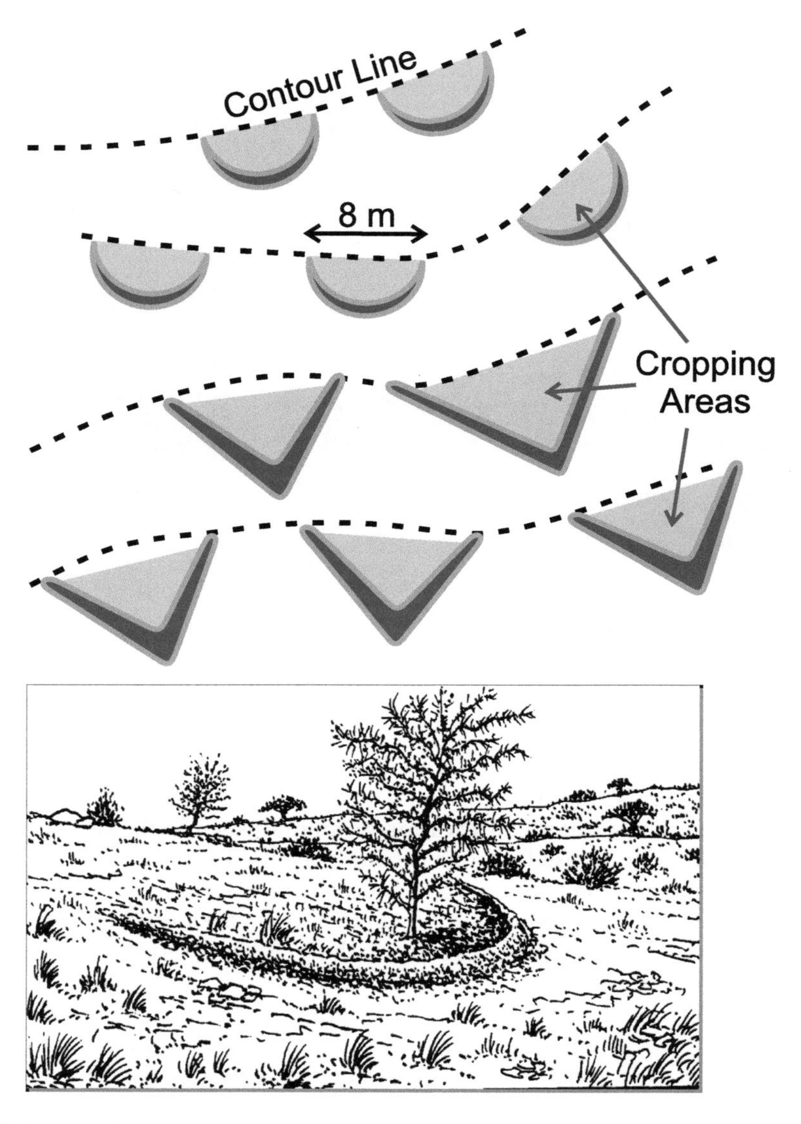

Figure 3.22 Layout of semicircular and triangular bunds. The areas between the cropping areas serve as catchment. Adapted from (Rocheleau *et al.*, 1988).

varies between 1 and 8 m; the bunds are 30 to 50 cm high. Crest or top width of bunds varies from 10 to 25 cm. Side slopes (horizontal: vertical) of bunds range from 1:1 to 3:1. Excavating soil at immediately uphill side of the bund location creates a slight depression, where runoff is intercepted and stored in the plant root zone.

The bunds can either be small and close to each other or larger and more widely spaced (Figure 3.23).

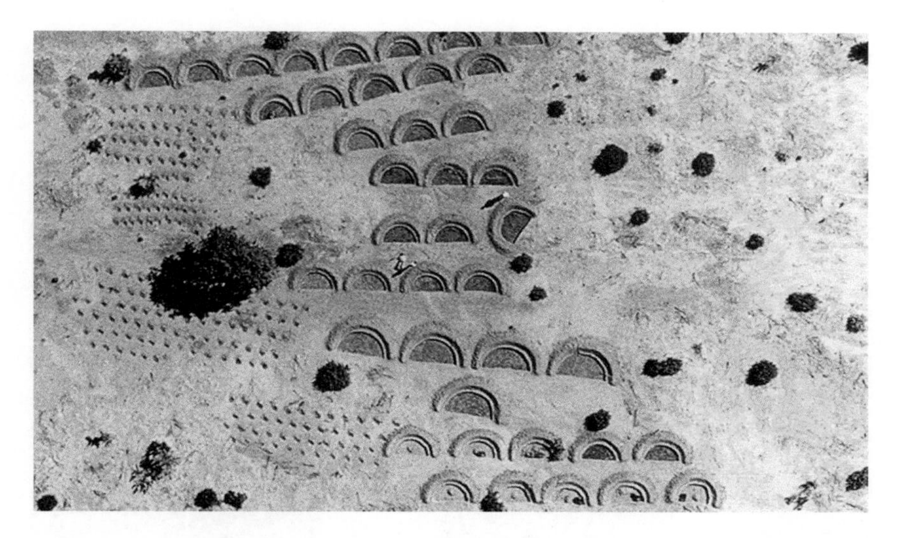

Figure 3.23 Micro-catchment water harvesting in Niger, including pits (left-hand side) and semicircular bunds. The larger semicircular bunds are 3 meters across (center); the smaller bunds are 2 meters across (lower two rows). Photo courtesy Deutsche Forschungsgemeinschaft (DFG). (*See color plate, page 243*).

Excavation of soil uphill from the bunds increases the slope and hence the runoff coefficient. This allows this technique to be used on even low slopes but it can be used on slopes up to 15%. This type of bund is mainly used for rangeland rehabilitation or fodder production (Figure 3.24), but can also be used for growing trees, shrubs, and in some cases field crops (e.g. sorghum) and vegetables (e.g. watermelons).

3.3.2.8 Eyebrow terraces

Eyebrow terraces are a form of semicircular bund that use stone to support the downhill side (Figure 3.25). The steeper the slope, the more the bunds have to be strengthened by stone material. The establishment and maintenance of this system are labor intensive.

These terraces have been reported from Ethiopia (Hurni, 1986; 1989), Tunisia (El-Amami, 1983), Jordan (Fardous *et al.*, 2004) and elsewhere.

3.3.2.9 Rectangular bunds

As the name suggests, rectangular bunds have bunds on three sides of the cropping area, with a catchment area on the open slope above the cultivated area (Figure 3.26). Shorter inner arms divide the cropping area into smaller basins to effectively marshal minor supplies of runoff (Van Dijk & Ahmed, 1993). Bunds are about 0.5 m tall with a base of about 2 m wide.

Figure 3.24 Semicircular micro-catchments with fodder bushes in northwest Syria. Photo courtesy D. Prinz/Karlsruhe University, Germany. (*See color plate, page 244*).

Figure 3.25 Fig trees growing in eyebrow terraces near Salloum in northwestern Egypt. Photo courtesy T. Oweis/ICARDA. (*See color plate, page 244*).

The *teras* system in Sudan is a typical example of a rectangular-bund water harvesting system (van Dijk & Reij, 1994). *Teras* systems are mainly used for the cultivation of sorghum but watermelons are sometimes planted on the bottom bunds where there is a relatively good supply of runoff. Maintenance is normally a modest

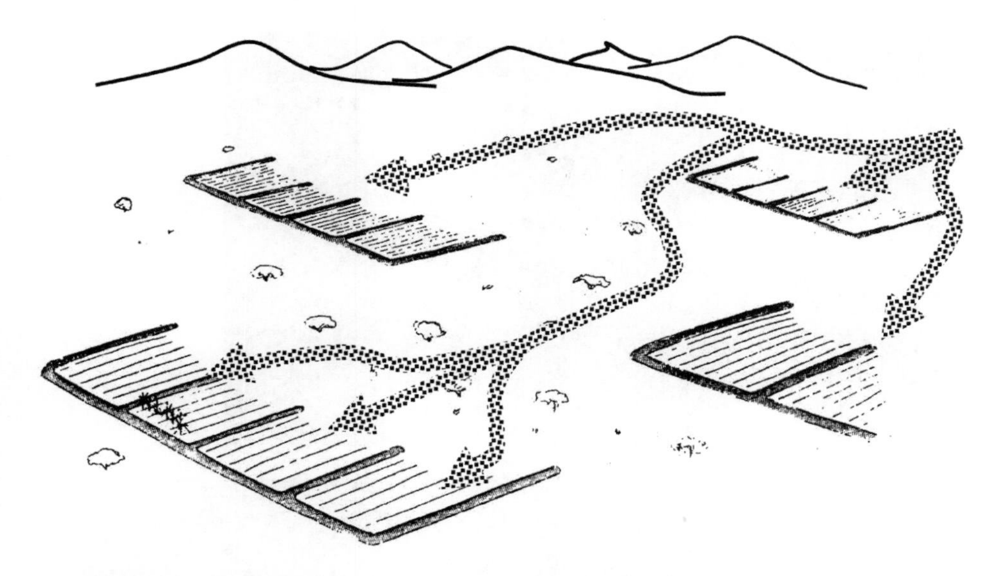

Figure 3.26 Rectangular bunds: the *teras* system in Kassala, northeast Sudan. Adapted from (Van Dijk & Ahmed, 1993).

task that can be carried out seasonally. This system cannot be combined with contour bunds for the rehabilitation of degraded lands, unlike the *zay* system.

3.3.2.10 Vallerani-type micro-catchments

Vallerani-type micro-catchments are constructed using a special plow mounted on tractor. This is used to construct a series of furrows and bunds along the contours of slopes (Figure 3.27). The bunds look like intermittent contour ridges. Being mechanized, this technique can be applied on a large scale (Figure 3.28). For example, about 10–15 ha were covered each day in the *badia* rangelands in Syria, creating over 5000 bunds or micro-basins each approximately 4–5 m long, 40 cm wide and 40 cm deep (ICARDA, 2006). Each micro-basin has a water-catchment capacity of about 600 liters. Establishing these micro-catchments costs about US$100/ha including planting, which is relatively high. However, this might be economical if large areas are to be covered (Antinori & Vallerani, 1994).

This type of micro-catchment is used for the establishment of forests, shelter belts, or agroforestry systems in arid and semi-arid areas. It can also be used for pasture improvement. In a project in Niger, this technique together with furrows constructed by plows was utilized for the cultivation of windbreaks. In another project at the Sinai desert, it was used for the cultivation of almond and olive trees (Malagnoux, 2009).

The Vallerani mechanized construction of micro-catchment implement for water harvesting was successfully tested by ICARDA project in the *badia* (dry rangeland) areas in Jordan, too. The plow was able to construct intermittent and continuous

(a)

(b) (c)

Figure 3.27 Vallerani-type micro-catchments: (a) the 'Wavy Dolphin Plow' used to construct the micro-catchments; (b) The plow in use; basins are constructed along the contour; (c) micro-catchments after plant establishment. Photos courtesy by (a) D. Prinz/Karlsruhe University, Germany; (b) ICARDA (2006); (c) Photo courtesy T. Oweis/ICARDA. (*See color plate, page 245*).

contour ridges, and could potentially be used to rehabilitate degraded rangelands. One major issue for large-scale implementation is the high cost and time required to manually identify contours for the plow to follow. A low-cost Contour Laser Guiding (CLG) system, with specifications that suit the contour ridging in undulating topographic conditions of dry rangelands was chosen, adapted, mounted, and

Figure 3.28 The Vallerani-type bunds look like an intermittent contour ridge. Mechanization facilitates construction of these bunds on a large scale. Photos courtesy T. Oweis/ICARDA. (*See color plate, page 246*).

tested, under actual field conditions. The system consisted mainly of a portable laser transmitter and a tractor-mounted receiver, connected to a guidance display panel. The system was field-tested on 95 ha of land where the system capacity was determined under different terrains, slopes (1–8%), and ridge spacings (4–12 m).

The easy adaptation and implementation of the CLG to the 'Vallerani' unit tripled the system capacity, improved efficiency and precision, and substantially reduced the cost of constructing micro-catchments for WH. The system is recommended for large-scale rangeland rehabilitation projects in the dry areas, not only in West Asia, but worldwide (Gammoh & Oweis, 2011).

3.4 MACRO-CATCHMENT WATER HARVESTING TECHNIQUES

3.4.1 Introduction

Macro-catchment water harvesting systems are characterized by collecting runoff water from a large natural catchment such as the slope of a mountain or hill. Catchments for these systems are often located outside the farm boundaries, where farmers have no control over them. The predominance of turbulent runoff and channel flow of the catchment water is a characteristic of macro-catchment systems, whereas micro-catchment systems are characterized by sheet or rill flow.

This macro-catchment method may be subdivided into two categories according to the nature of the catchment and the way runoff water is flowing and/or is transported to the target: long-slope systems and floodwater systems.

In the long-slope systems runoff water usually flows on natural sloping land with short channels, if any. The catchment area is somewhat defined and not far from the target area, hence travel time of runoff water is usually short (Mzirai & Tumbo, 2010). In floodwater harvesting runoff water is usually generated from a remote, vast and ill-defined catchment and transported via a long and well-defined channel (*wadi*). Floodwater systems may be further subdivided into two classes according to the location of the target: *wadi*-bed if the target is inside the *wadi*'s cross section; and off-*wadi*-bed if the target is located outside the cross section of the *wadi* channel. The latter involves some means of diverting water outside the *wadi* channel.

Generally, the proportion of runoff captured per unit area of catchment is much lower in macro-catchments than in micro-catchments and ranges from 10 to 50% of annual rainfall. Water is often stored in surface or subsurface reservoirs, but may also be stored in the soil profile for direct use by crops. In some cases, water is stored in aquifers as a recharge system. The cropping area is either terraced on gentle slopes or located on flat terrain (Figure 3.29).

In many locations, a mixture of several macro-catchment types or a combination of macro- and micro-catchment techniques can be found (Figure 3.30).

3.4.2 Long-slope water harvesting

3.4.2.1 Hillside conduit systems

In hillside conduit systems rainwater running down hill is directed by small conduit channels to fields at the foot of the hills or mountains (Evenari *et al.*, 1982; Klemm, 1990) (Figure 3.31). They are found in many semi-arid hilly or mountainous regions. In Pakistan, this system is called *sylaba* or *sailaba*. These systems have been the backbone of agriculturally based civilizations in the Middle East and North Africa for millennia (Prinz, 1994) and has been used by native Americans in North America, where runoff water from sandstone outcrops was used by the Navajo to grow maize and squash under annual rainfall of 300 to 400 mm (Figure 3.32) (Vivian, 1974).

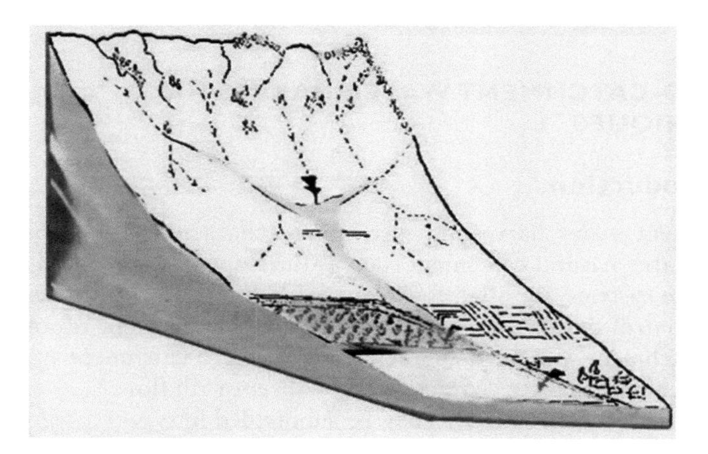

Figure 3.29 Schematic diagram of a typical macro-catchment water harvesting system. (Oweis *et al.*, 2001).

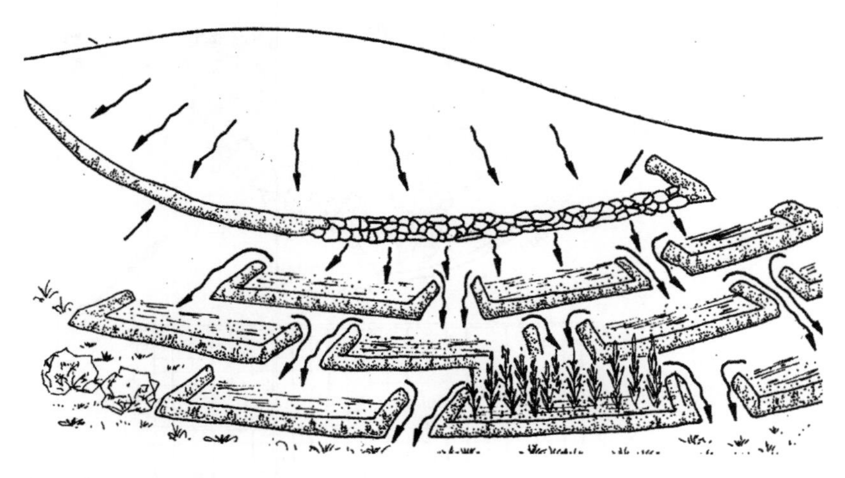

Figure 3.30 Combining macro- and micro-catchment systems in Mauritania. (Tabor & Djiby, 1987).

Figure 3.31 A hillside conduit water harvesting system intercepts, collects, and directs runoff water down to the target area in the valley or at the foot of the hill. Reconstruction of a Nabataean farm in the Negev. Photo courtesy D. Prinz/Karlsruhe University, Germany.

In these systems, fields are leveled and surrounded by levees with a spillway to drain excess water to fields downstream. When all fields in the series are filled by water, water is allowed to join the *wadi*. Distribution basins can be constructed to allow the use of several feeder canals (see Figure 3.33 and Box 3.2). This is an ideal system to utilize the runoff from bare or sparsely vegetated hilly or mountainous areas (Mzirai & Tumbo, 2010).

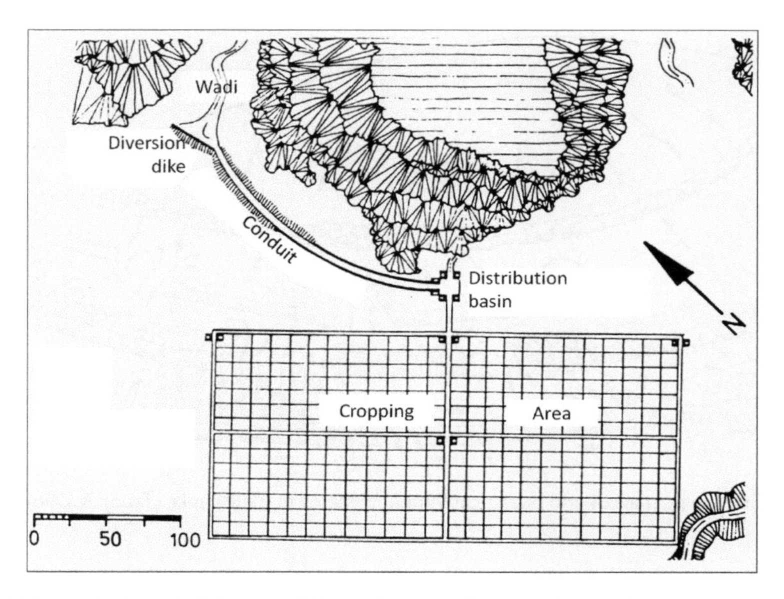

Figure 3.32 Navajo Indians in Arizona, USA, used a conduit water harvesting system to grow maize and squash. (Hudson, 1987 after Vivian, 1974).

Figure 3.33 Hillside conduit system in Mali: Flooded fields with a diversion. (Klemm, 1990). *(See color plate, page 246).*

Hillside conduit schemes require proper design, high labor input, and probably the assistance of an engineer. The slopes of the conduits should be sufficient to prevent sediment settling; alternatively sediment must be cleared after heavy rainstorms. Fields need to be leveled and spillways constructed at the correct height to ensure uniform distribution over the fields and equity among beneficiaries. When large amounts of rainfall are harvested regularly, bunds and spillways may be used to control the amount of water delivered to various cropped areas located down the slope.

This technique can be applied for the cultivation of many crops and fruit trees, especially those that can stand some days of waterlogging. Sorghum, for example, which is tolerant of waterlogging and is thus well suited to this system, whereas maize, which has got a low tolerance towards waterlogging, is not. A system of bunds and spillways can, however, be used to control the amount of water delivered to various cropped areas, which would allow sorghum to be planted on the upper plots and drought-resistant millets to be planted on the lowest plots.

Box 3.2 Case study: Macro-catchment water harvesting in Mali, West Africa. (See Box 8.3 for further information)

Problem analysis: 570 mm annual precipitation; >95% of rain concentrating in 5 months; high population growth (3%/yr); small holdings; subsistence cropping.

Goals: Sustaining life for a rural community under conditions of desertification, decreasing rainfall, and growing population.

Framework conditions: Initiated by a German engineer, the work was carried out by 30 to 50 farmers of the region of Yelimane, Kayes Province, and western Mali.

Technical solutions: Macro-catchment water harvesting using the hillside conduit system. Runoff from an 81-ha catchment is directed to a 3.3-ha terraced cropping area (Figure B3.2.1) and distributed there.

Achievements: Yields were significantly increased and cropping risk reduced; desertification in that region halted, farmers became more self-confident.

In the first phase, stream gauges and a meteorological station were established, soil samples were taken, and a 1:10 000 topographical map was drawn, based on aerial photographs. Land use was recorded and farmers were interviewed.

Figure B3.2.1 The hillside conduit system in Kanguessanou, western Mali. The distribution basin is located at the base of the hill (foreground). (Klemm, 1990).

In the second, 2-year phase, the water harvesting system was designed and constructed, almost exclusively using local material and the labor of the beneficiaries.

The soils in the valley are fluvisols and regosols; the fields have been terraced (Figure B3.2.2); the hydraulic structures are laid out for 10 cm field flooding height.

The runoff coefficient of the catchment areas was between 0.2 and 0.3; a catchment:cropping area ratio of 10:1 was regarded as optimal to satisfy the demand of the staple crops (sorghum, cowpea, maize, peanuts). If rainfall was higher than demand, a drainage channel evacuated surplus water (Figure B3.2.3).

Figure B3.2.2 The same location during the rainy season. The 3.3-ha cropping area shows good crop growth. (Klemm, 1990).

Figure B3.2.3 Drainage channel to evacuate surplus water after filling all fields. (Klemm, 1990).

3.4.2.2 *Limans*

Limans are single structures at the foot of long slopes, consisting of a bund of 1–3-m high around a cropping area (Figure 3.34). They are not located in valleys as *jessour* systems are, and they are not arranged in rows as is the case with large semicircular bund systems. The term 'liman' was selected for this system because after a runoff-producing storm, the cultivated area looks like a lake or a pool or liman (Figure 3.35). The Tunisian term for it is '*tabia*', a term which is also used for large semicircular bunds (Chapter 3.4.2.3).
This technique is mainly used in regions of very low precipitation and with very few rainfall events per year. The size of the cropping area varies from 0.1 to 0.5 ha, while the catchment area may be up to 200 ha.

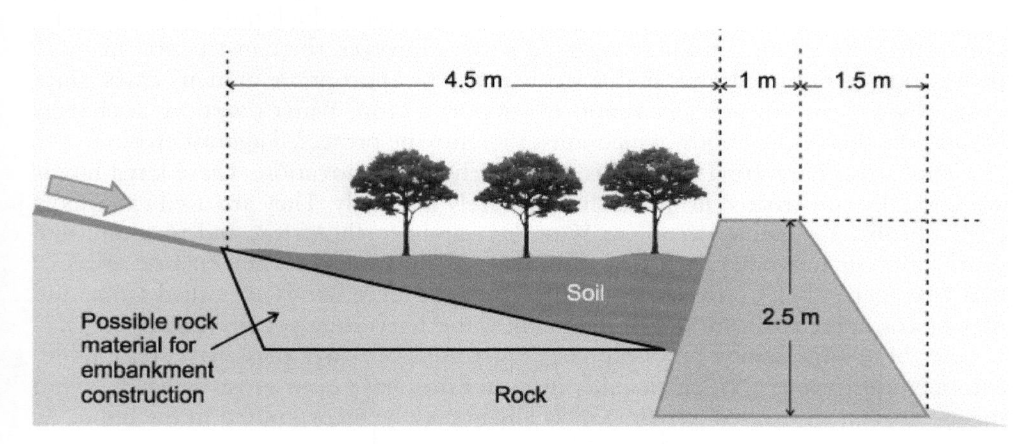

Figure 3.34 Typical cross-section of a liman. The embankment may be constructed from masonry or from compacted clay-rich soil. Rock for the embankment may be dug from the site itself to increase the space for deposited soil. Figure provided by authors.

Figure 3.35 A flooded liman. The trees are four years old. Photo courtesy H. Loewenstein/Ben Gurion University, Beersheva, Israel.

Limans can be planted with trees that are tolerant of waterlogging yet able to withstand months of drought. Examples include *Eucalyptus occidentalis* and *Acacia salicina*. The eucalyptus can be planted at a density of up to 700 per hectare (Bruins *et al.*, 1986). Once the trees have been established, the cultivated area can be used for grazing. In rare cases, salinization has been observed in this system.

3.4.2.3 Large semicircular or trapezoidal bunds

These structures consist of large semicircular, trapezoidal or V-shaped earthen bunds facing up the slope (Figure 3.36). The bund traps runoff from the catchment area above the structure. Crops are planted when the water trapped in the enclosed area subsides. The system is best suited to slopes of 1–3%. The distance between the tips of each bund may range between 10 and 100 m or more; the bunds are 1–2 m tall. Often, they are aligned in long staggered rows. However, the construction of more than two rows of these bunds in one site may not be appropriate in many cases, since lower rows receive insufficient runoff to support a crop. Water overflow discharges around the tips of the bunds, which must therefore be protected against erosion.

This water harvesting method is a relatively new innovation. These large bunds are normally constructed using machinery, rarely manually. They are used to support trees, shrubs, and annual crops in West Asia and North Africa, and sorghum and pearl millet in sub-Saharan Africa (Van Dijk & Reij, 1994). They enclose areas of 0.1–1 ha, depending on the slope. In Tunisia, these large bunds are called *tabia*, and may be considered as a variant of the liman water harvesting system (Figure 3.37).

As these large bunds can store large quantities of water they may break under extreme rainstorm events, particularly just after they have been constructed. To avoid this problem a controlled overflow mechanism may be incorporated in the bund. As the systems are not traditional, adoption can be difficult.

3.4.2.4 Cultivated tanks/reservoirs and hafairs

Tanks are usually earthen reservoirs dug in the ground in gently sloping areas that receive runoff water either by diversion from *wadis* or from a large catchment area. They are known as 'Roman ponds' in parts of North Africa, where they are usually

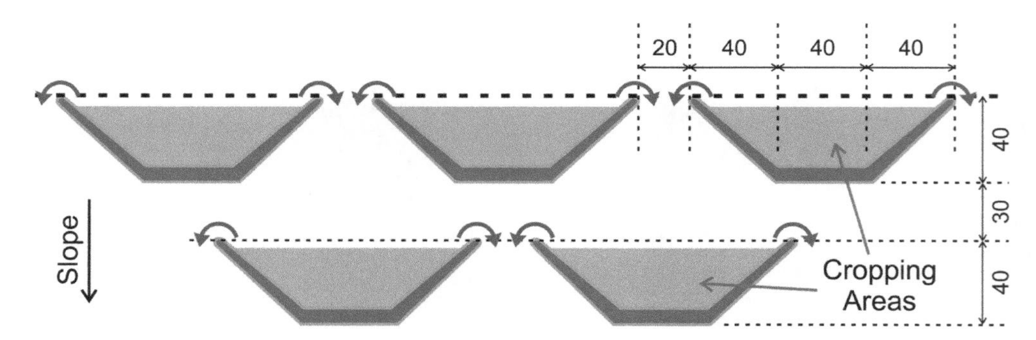

Figure 3.36 Layout of large trapezoidal bunds. Dimensions are in meters. Adapted from (Critchley & Siegert, 1991).

Figure 3.37 Typical *tabia* in southern Tunisia. Photo courtesy T. Oweis/ICARDA. (*See color plate, page 247*).

built with stone walls. The capacity of these tanks ranges from a few thousand cubic meters, in which case they are called *hafair*, to tens of thousands of cubic meters. Cultivated tanks are very common in India, where they provide irrigation water to over 3 million hectares of cultivated land. In the WANA region, especially Sudan, Jordan, and Syria, smaller tanks are more common and are mostly used to store water for human and animal consumption (Figure 3.38).

Several problems are associated with tanks and reservoirs. The water may become stagnant and polluted, provides a breeding ground for insects, especially mosquitoes, and can become a source of disease. As these reservoirs usually have no protection around them, there is a risk of people or animals falling into them and drowning. Large amount of water is lost through seepage and evaporation. Several improvements, including fencing, lining to reduce seepage losses, and settling basins to capture sediment load, have been introduced to overcome these limitations (see Chapter 7 for more details).

3.4.3 Floodwater harvesting systems

The main characteristics of floodwater harvesting systems, also called 'spate irrigation', are:

- Large, distant catchments (some several kilometers from the target cropping area)
- Turbulent flow of water in channels

Figure 3.38 A *hafair* – small reservoir – in Sudan. Photo courtesy T. Oweis/ICARDA.

- Water is either stored and then diverted or spread within the stream bed
- Solid hydraulic structures
- Provision for the removal of excess water.

Among the most important problems associated with these systems are water rights and the allocation of water between the catchment and the cultivated areas and among users upstream and downstream of the watershed. These are best addressed by planning the water harvesting interventions within an integrated watershed development approach. An excellent overview over all aspects of floodwater harvesting is given in the 'FAO guidelines on spate irrigation' (FAO, 2010). Van Steenbergen *et al.* (2011) may serve as a good example how to assess and develop the floodwater harvesting potential of a country. A 'Spate Irrigation Network' (www.spate-irrigation.org) serves a.o. the exchange of experiences among professionals and practioners.

3.4.3.1 *Wadi-bed water harvesting systems*

In *wadi*-bed water harvesting systems the bed of the *wadi* is used to store the water either on the surface by blocking the water flow or in the soil profile by slowing the flow and allowing it to infiltrate in the soil. They involve complex systems of dams and distribution networks.

Farmers who have a *wadi* passing through their land can build a small dam in a suitable location to store some or all of the runoff water in the *wadi* (Figure 3.39). More details on storing harvested water are given in Chapter 7.

This technique is very common in *wadi* beds with mild slopes. Due to slow water velocity, eroded sediments usually settle in the *wadi* bed and create good

Figure 3.39 A *wadi*-bed water harvesting system in Tunisia. Photo courtesy T. Oweis/ICARDA. (*See color plate, page 247*).

agricultural land. This may occur naturally or by constructing small dams or dikes across the *wadi* to reduce the flow velocity and encourage sediment to settle. These walls should be made of stone or gabions, and should not be more than one meter tall (Figure 3.40). The top of the wall should be horizontal in order to create level land behind it and allow excess water to overflow along the entire top length. The distance between walls along the *wadi* bed depends on the slope of the *wadi* bed and the height of the wall.

Crops commonly grown in *wadi* beds include fruit trees such as fig, olive, date palm, and other high value crops since the soil in the *wadi* bed is usually fertile and water is reasonably guaranteed. The main problem with this type of water harvesting is the cost involved in building and maintaining the crossing walls. Another problem that appeared recently in some parts of WANA is that increasing human activity in catchment areas is resulting in less runoff water reaching the *wadis* depriving

Figure 3.40 Stone walls constructed using gabions across a *wadi*-bed in Marsa Matruh area in north-west Egypt. Photo courtesy T. Oweis/ICARDA. (*See color plate, page 248*).

downstream cultivation areas of water. This indicates the need for integrated watershed development.

A variant of *wadi*-bed water harvesting is practiced in steeper *wadis* in southern Tunisia, where the system is known as *jessour*. In this system the walls (*tabia*) are usually built to a height of 1–2 m initially and are raised as sediment accumulates behind them (Figure 3.41). They are made of earth, stones or both, but always have a spillway, usually made of stone. The top of the *tabia* is normally 30 to 40 cm above the soil surface level immediately upstream the *tabia*.

Usually, there are series of *jessour* along the *wadi*. These systems require maintenance to stay in good shape. Since the importance of these systems for food production has declined recently, attention to maintenance has also declined and many systems are losing their function.

The dimensions and design of these structures greatly vary depending on the site geometry and water flow rates. They are usually keyed in the *wadi*-bed by digging a foundation trench to prevent them from being undermined by runoff. Labor and stone requirements are high. It may be necessary to transport stones to the site if not enough stone is available locally.

Similar systems are employed in China, where it is known as 'warping' (Figure 3.42). In this system the valley bottom is dissected by dams between 100 m

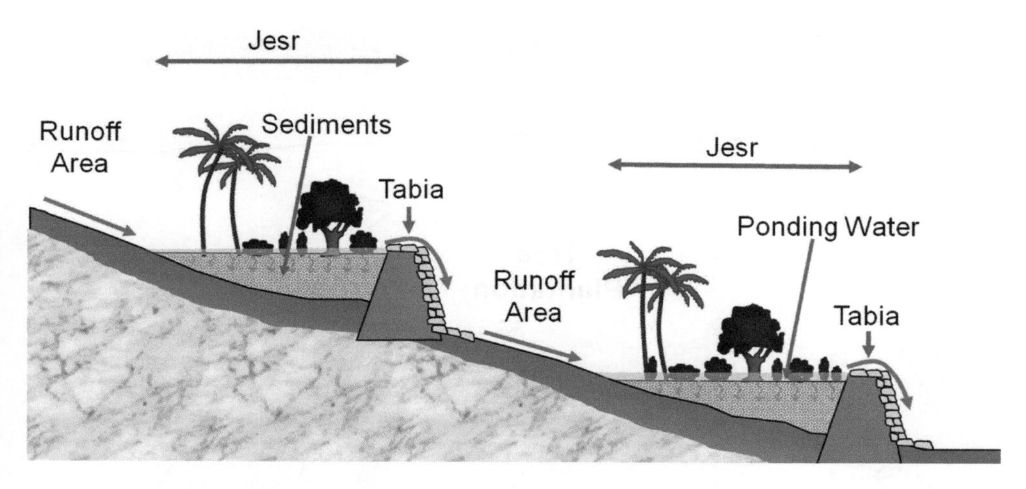

Figure 3.41 Schematic drawing of a series of *jessour (jesr = singular)* built along a steep *wadi* bed.

Figure 3.42 An example of warping from the loess plateau in central China. (Yuanhong & Qiang, 2001). (*See color plate, page 248*).

and several kilometers apart. Sediment eroded from the adjacent cultivated areas on the loess plateau is spilled into the segments, filling them up over the course of time. The newly reclaimed areas are highly fertile and are intensively cropped.

3.4.3.2 *Off-wadi systems*

In off-*wadi* systems structures force the *wadi* water to leave its natural course and direct it to nearby areas suitable for agriculture (Figure 3.43). This technique is also called 'floodwater diversion'. Similar structures may also be used to collect rainwater

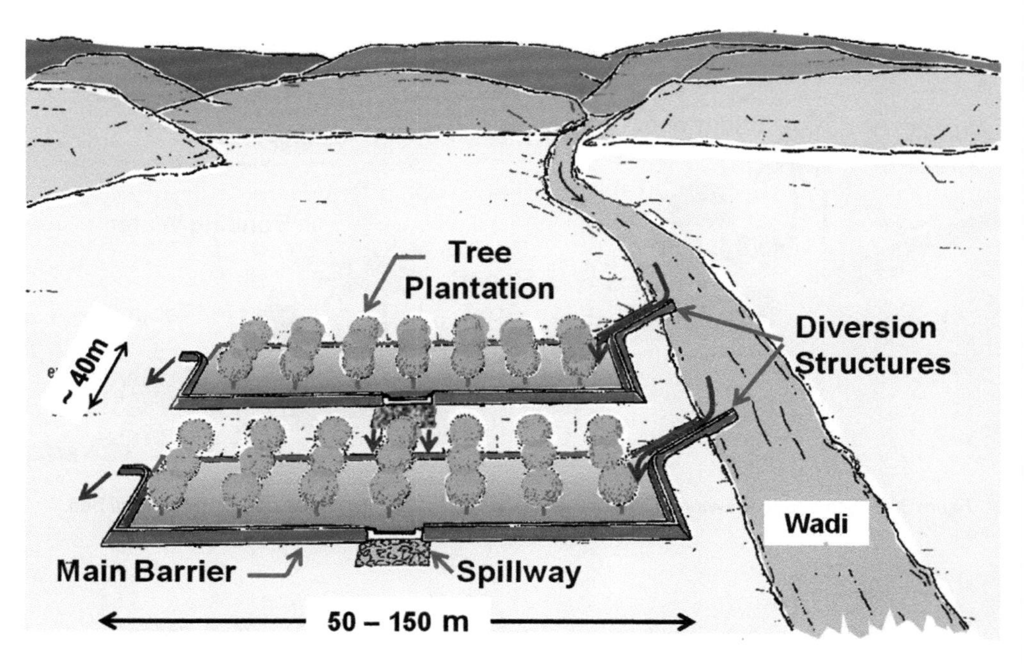

Figure 3.43 Example of floodwater diversion as applied in Tunisia. Adapted from (GTZ/DGF, 1993).

from catchments outside the *wadi* bed, such in the cases of tanks and *hafairs*. In this system water is stored only in the root zone of the crops, i.e. it supplements rainfall.

Water diversion requires relatively uniform land with a low slope. Agricultural land may be graded and divided into basins by levees to allow enough water to be stored for the season. Soils should be deep with good water-holding capacity.

The key to the success of this system is the construction of the diversion structure and the canals conveying the water, which may require the expertise of an engineer. The diversion structure must be strong enough to resist the flow of the *wadi* and at an elevation appropriate to divert the required portion of the flow. Various materials have been used to build diversion structures, including stone and concrete, but the most durable material is gabions.

One important point to consider is that the slope of the canal should allow a flow fast enough to prevent the accumulation of sediments near the structure; otherwise it will block and require frequent, expensive maintenance.

3.5 HARVESTING WATER FOR ANIMAL CONSUMPTION

3.5.1 Traditional techniques

Desert dwellers usually harvest surface runoff for animal and human consumption by redirecting water running downhill slopes into cisterns (Ali *et al.*, 2009; see also Figure 7.16). In many regions of the world, this water harvesting is still practiced, but with some modifications to suit the end users. In some semi-arid African countries where pastoralism is the major occupation of the people, traditional ways of collecting water for animal consumption include excavated cisterns, *hafairs*, small dams, and natural water holes (see Chapter 7 for more details).

3.5.2 Modern techniques

In modern times, most systems used to harvest water for animal consumption have concentrated on modifying the catchment area to increase the total amount of water harvested and designing better storage structures (Figure 3.44). In Australia, for instance, runoff from roads is used for watering livestock. Another technique to enlarge the volume of runoff for underground storage is the construction of up to 20 m long wings on both sides of a cistern. The collecting wings, made of concrete or masonry, should have a solid base and a height above ground of 30–50 cm. Common treatment techniques are discussed in detail in chapter 4.

3.6 CONTAMINATION CONCERNS

The major problem in harvesting surface runoff for domestic and animal consumption is the possible contamination of the harvested water. Since the surfaces from which the water is harvested are exposed throughout the year, they can be contaminated by animal and human dung, dust, insects, and birds. Chemicals used to treat catchment surfaces are sometimes washed down the storage area by the runoff. The issues of contamination are elaborated in Chapter 10.

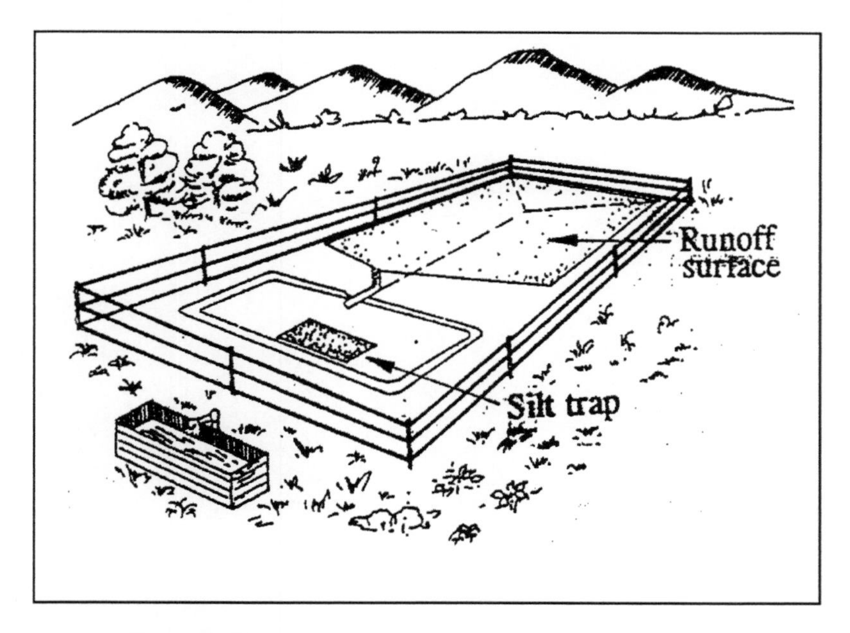

Figure 3.44 A system used to harvest rainwater for animal consumption in Western Australia. (FAO, 1977).

Chapter 4

Runoff inducement methods

4.1 INTRODUCTION

The term 'water harvesting' is used to describe the process by which water is collected from an area that may have been modified or treated to increase precipitation runoff and is stored for later use. This chapter describes some of the methods and materials used to induce runoff in agricultural systems. Some general ideas and concepts are presented on runoff inducement techniques (such as surface modification, cover sheets, and soil surface treatment) that are feasible for use in arid and semi-arid regions of the world. The advantages and disadvantages, cost, and conditions favoring each method are discussed.

4.2 METHODS OF IMPROVING RUNOFF

Often the catchment area needs to be modified to increase runoff (see, for example, Box 4.1). This can be done by:

- modifying the topography or soil surface;
- modifying the soil; and
- covering the surface with an impermeable layer.

There is no one technique or method that is best in all situations. The best techniques to use vary depending on topography, soil condition, storage devices, labor, availability of treating/covering materials, and intended use of the water harvested. The cost of alternative water sources and the importance of water supply determine the costs which can be justified.

The total cost for preparing a catchment area is composed of two main items: the cost of materials and cost of labor. Some materials and installation techniques are labor-intensive but have relatively low capital costs. This type of techniques may be suitable for areas where labor is cheap. Other approaches may have high capital costs but require a minimum of labor, e.g. mechanized compaction (Figure 4.1). Such techniques may be appropriate in areas with high labor costs. Usually, the water harvesting systems used in runoff farming are constructed from materials that are cheap, locally available, and easily handled.

Figure 4.1 Using a roller to compact runoff strips between crop strips has high capital cost but low labor requirement. Photo courtesy T. Oweis/ICARDA. (*See color plate, page 249*).

4.2.1 Creating shallow channels

In long-slope water harvesting systems the runoff water yield can be increased substantially by creating shallow channels within the catchment area. Depending on local conditions, the work involved can be done manually or by heavy machinery. Special care is needed to avoid soil erosion within those channels. The construction of small bunds perpendicular to the direction of flow slows down the running water, promotes sedimentation and reduces the erosion risk.

4.2.2 Clearing the catchment

Clearing rocks and vegetation usually increases runoff (Figure 4.2). Only some of the gravel, stones, and vegetation need be cleared, with little modification to the topography or surface structure. Clearing the catchment in this way can be a very economical way to harvest rainwater in arid lands if erosion is limited and low-cost hillside land is available. However, if erosion is severe, soil conservation measures have to be selected that do not significantly reduce runoff water yield.

4.2.3 Smoothing the soil surface

The soil surface may be smoothed by removing small obstructions such as ridges and furrows across the contour of the land. In this method, small amounts of topographic modification are required. This may require considerable amounts of labor or the use of machinery, depending on the topography and soil conditions. Smoothing the soil surface as sole method applied, may have low runoff efficiencies. Runoff efficiency can be improved by constructing a system of ditches and ridges, arranged in a fish-bone style, on suitable slopes. These treatments are effective on suitable soil types and with appropriate topography.

(a)

(b)

Figure 4.2 Clearing fields to improve water harvesting in Jordan. The stone collected can be used to build stone bunds (a) or eyebrow terraces (b) (Ali *et al.*, 2006).

4.2.4 Compacting the soil surface

Compacting the soil surface can increase the runoff. The slopes are graded and compacted manually or mechanically. For manual compaction, a hand hammer may be used; mechanical compaction requires a tractor and rubber-tired roller or other compacting machinery, depending on field conditions and the area to be compacted (Figure 4.3).

Compacting and smoothing methods have been used successfully in 'roaded catchments' in Western Australia (Figure 4.4). The area has an average annual rainfall of 500 mm, falling during seven winter months. The surface layers of the soil are sandy, while the subsoil is clayey. The sand is moved into rows, exposing the clay. This is then shaped and spread to cover the whole surface. The ridges discharge induced runoff water into a channel which conducts it to a tank with a capacity of 3000 m^3.

Figure 4.3 Smoothing and compacting the soil surface at ICARDA's Field Station. Photo courtesy T. Oweis/ICARDA.

Figure 4.4 A roaded catchment with a reservoir in Australia (Western Australia, Central Regions Development Advisory Committee, 1992). (*See color plate, page 249*).

The major advantage of this method is that the system uses the existing soil and can be built with readily available equipment. Compacting and smoothing the steep road surfaces is most important, here achieved by tractors and rubber-tired rollers. This method of runoff inducement may be used in other arid and semi-arid areas having similar soil and topographic conditions. However, the high capital cost of the technique makes it unsuitable for many developing countries.

4.2.5 Surface sealing

Surface sealing involves chemical treatments either sprayed onto the catchment area or mixed into the soil surface to reduce or stop water infiltration (Figure 4.5). Many types of materials have been tested. Unfortunately, many of them work only on specific soil types and are not successful for long-term use. Light soils with high infiltration rate do not produce much runoff. This is a major problem in many sandy areas of WANA where water harvesting is very much needed. The search continues for practical, low-cost, and environmentally friendly materials that can be used for this purpose.

Sodium salts cause clay in the soil to disperse, swell, or break down into small particles, increasing runoff. They show promise as a soil sealant because of their low cost, ready availability, and retardation of weed growth. However, soil erosion might be a potential problem with this treatment. Negative effects on plant growth by this kind of treatment have not been observed.

The treatment consists of mixing a water-soluble sodium-based salt (e.g. NaCl) into the top 2 cm of soil at a rate of about 1 t/ha. The catchment area is then wetted and compacted to a firm, smooth surface. This treatment requires a soil containing 20% or more of a kaolinite- or illite-type clay. The sodium salt disperses the clay which plugs the soil pores and reduces the rate of water permeability (Dutt, 1981). Some experiments show that some of these materials work perfectly as water repellent but may enhance the weathering ability of the soil. However, more information is needed to predict the level of performance of this type of treatment.

A second type of soil surface modification treatment is the application of water-repellent chemicals. These materials when applied to the catchment area create a

Figure 4.5 A modern rainwater harvesting catchment site in Mexico. The soil surface has been smoothed and compacted using a tractor and a rubber-tired roller and salt has been mixed into the soil surface to create optimal conditions for high runoff. The water not used by the plants is collected at the foot of the slope and is used for drip irrigation of the orchard during the dry season. Photo courtesy G.W. Frasier, USDA.

hydrophobic or water-repellent soil surface. This treatment does not change the porosity of the soil, but instead changes the surface-tension characteristics between the water and soil particles. One of the simplest water-repellent chemicals is sodium silanolate. This is applied in a water solution and forms a water-repellent layer 1–2 cm deep with an effective life of 3–5 years. The treatment does not provide any soil stabilization; hence wind and water erosion can be a problem. It is not suited for soils containing more than 15% clay.

Another water-repellent treatment consists of spraying molten, refined, low-melting point paraffin wax onto the prepared soil surface. The wax is initially deposited as a thin layer on the surface. As the sun heats the surface, the wax partially melts and moves deeper into the soil, coating each individual soil particle with a thin wax layer and rendering the soil water repellent. This treatment is best suited to soil containing less than 20% clay and on catchment sites where the soil temperature will exceed the melting point of the wax during part of the year (Frasier, 1980). Wax-treated plots yield an average of 90% of the rainfall as runoff, compared with 30–40% from untreated plots. However, the paraffin wax does not provide significant soil stabilization and the treatment is susceptible to water and wind erosion.

To overcome the difficulties of using paraffin wax in inducing runoff, the wax may be emulsified by using low-cost additives. The emulsified wax can be applied easily to catchment plots using a small sprayer (Figure 4.6). The use of the wax may triple the amount of runoff from small plots.

4.2.6 Impermeable coverings

Instead of making the soil itself the water-shedding surface, it may be better in some situations to cover it with a waterproof layer. Most types of plastic and other thin sheeting materials have been investigated as potential soil coverings for water harvesting catchments, including thin plastic films, butyl rubber, asphalt membranes, and highway surfaces. Bitumen or asphalt are best suited to fine sandy soils, but have an effective life of only 2–5 years (Laing, 1981). Unfortunately, thin film coverings are susceptible to damage by wind and/or sunlight.

Figure 4.6 Applying emulsified paraffin wax on a micro-catchment near Aleppo, Syria, to induce surface runoff. Photo courtesy T. Oweis/ICARDA.

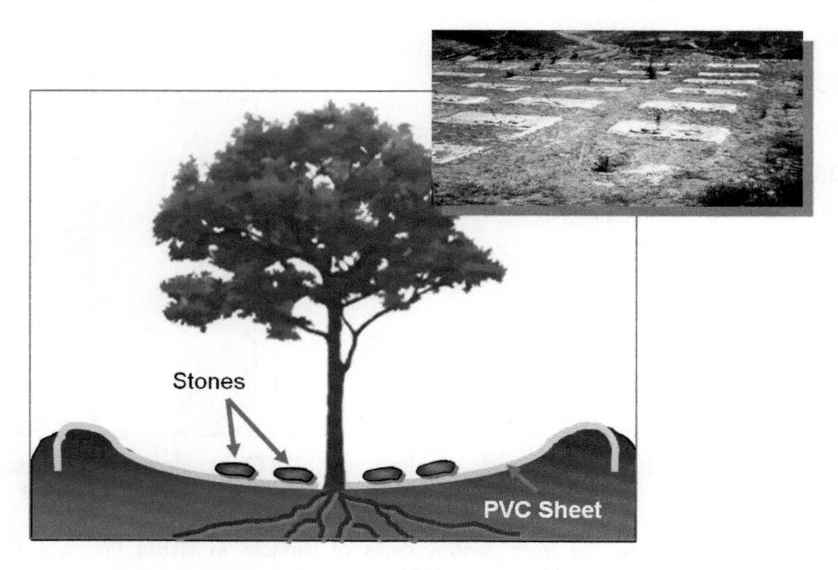

Figure 4.7 Partial covering of the ground with plastic sheets in a plantation in Gansu, China. Photo courtesy Zhu Qiang/Gansu Research Institute for Water Conservancy, PRC.

The partial covering of the ground with plastic sheets, e.g., around trees, contributes to evaporation reduction and increase in rainwater collection (Figure 4.7).

Box 4.1 Water harvesting from various types of surfaces in Gansu Province, China.

Framework conditions: Farmers in the Loess Plateau of Gansu Province face not only very difficult topographic conditions and substantial soil erosion, but have to live with barely 330 mm annual rainfall (average) that is unevenly distributed: 50–70% of the rain falls in July–September; only 19–24% falls during the main crop growth period. Per capita water availability is only 230 m^3, compared with an average of 2300 m^3/person in China and the world average of 7000 m^3/person.

Technical solutions: The farmers practice an efficient type of dryland agriculture, but cannot survive without rainwater harvesting. They collect rainwater from various types of surfaces and store it underground in cisterns.

The farmers have created a so-called 'one-fourth system': The land is divided into four parts and one part is covered with (preferably UV resistant) plastic film to collect water for the other three parts. Each of the parts is used as the catchment in turn. The rainwater collection efficiency is raised from only 0.05–0.08 on the natural slope to about 0.9 from land covered with plastic sheets. The runoff water is directed to cisterns. Water is also collected from roads, with a runoff efficiency of 0.75 for asphalt-paved highways. Land used for plantations is covered with plastic sheets between the rows of trees to suppress evaporation and to collect surplus water. (Figure B4.1.1).

Achievements: Partial success has been achieved with some of these materials by bonding them to the soil surface with an asphaltic compound. Impermeable coverings have been found to be very successful on porous or unstable soils in particular. Other methods have proved to be too costly.

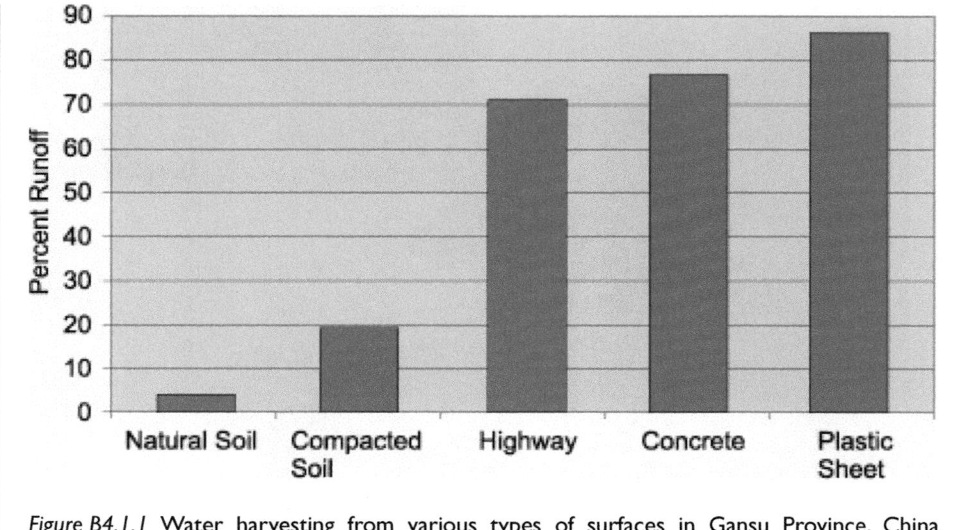

Figure B4.1.1 Water harvesting from various types of surfaces in Gansu Province, China (% runoff). (Qiang & Yuanhong, 2001).

One of the simpler techniques for utilizing low-cost sheets of plastic or roofing tar paper is to place a shallow layer of clean gravel on the sheeting after it has been positioned on the catchment surface. The sheets of plastic/the tar paper provides the waterproof membrane and the gravel protects it from wind and sun. A good catchment surface can also be made by covering asphalt with a better-quality film with a gravel layer on top. The asphalt bonds the film to the catchment surface, while the film protects the asphalt from oxidation. This treatment requires periodic maintenance to ensure the sheeting remains covered with gravel. Runoff is essentially 100% of all precipitation in excess of 2 mm. These catchments, if properly constructed and maintained, can last for 20 years. Wind-blown dust trapped in the gravel layer is a potential seedbed for plants, which could negatively affect runoff efficiency. The treatment is relatively inexpensive, if clean gravel is readily available (Cluff, 1975). Experiments conducted in the USA on the use of asphalt as catchment cover demonstrated the following:

- Strong and durable catchments for runoff inducement and water harvesting can be constructed by spraying asphalt on the surface of loamy sand and sandy loam soils.
- The larger the catchment size, the lower the cost of construction per unit area.
- In areas with high solar radiation and low precipitation, runoff from asphalt catchments was colored by asphalt oxidation agents. This colored water was consumed by cattle without problem.

An effective treatment used for supplying water for wildlife and irrigation is the asphalt–fabric membrane. In this system random-weave fiberglass matting or a synthetic polyester engineering filter fabric is unrolled on the prepared catchment surface and saturated with an asphalt emulsion. Three to 10 days later, a final asphaltic emulsion seal coat is brushed on the membrane. These membranes are relatively resistant to damage by wind, animals, and weathering.

Many conventional construction materials such as concrete, sheet metal, or artificial rubber sheeting can be used on water harvesting catchments (Box 4.2). These materials are relatively expensive, but when properly installed and maintained have an effective life up to 20 years. They are useful where gravel is readily available and maximum runoff is not required.

Box 4.2 Utilizing slopes covered with plastic sheets for rainwater harvesting.

According to Chinese law, slopes of greater than 47% inclination may not be used for agricultural purposes, except for trees and bushes. In dry areas of the country, these slopes could be utilized for harvesting rainwater by placing black (preferably UV-resistant) plastic sheets on the ground. The edges of these sheets are buried in the ground; the central line is slightly deeper than the remaining portions. Some stones are placed in the central line to stabilize the sheets and to prevent wind damage. Fruit trees are planted alongside the plastic sheets to serve the same purpose.

Runoff is collected in a basin at the foot of the slope, from where the water is pumped (by hand or diesel pump) to a small tank. This tank is placed about 2 m above the ground to allow the application of a simple drip irrigation system for cultivation of high-value horticultural crops. (Figures B4.2.1 and 4.2.2).

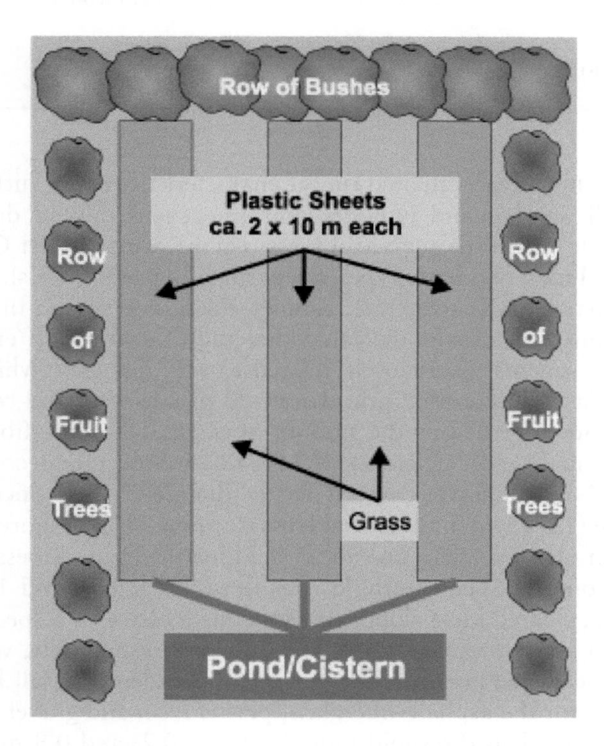

Figure B4.2.1 Utilizing slopes covered with plastic sheets for rainwater harvesting (Prinz, 2002a).

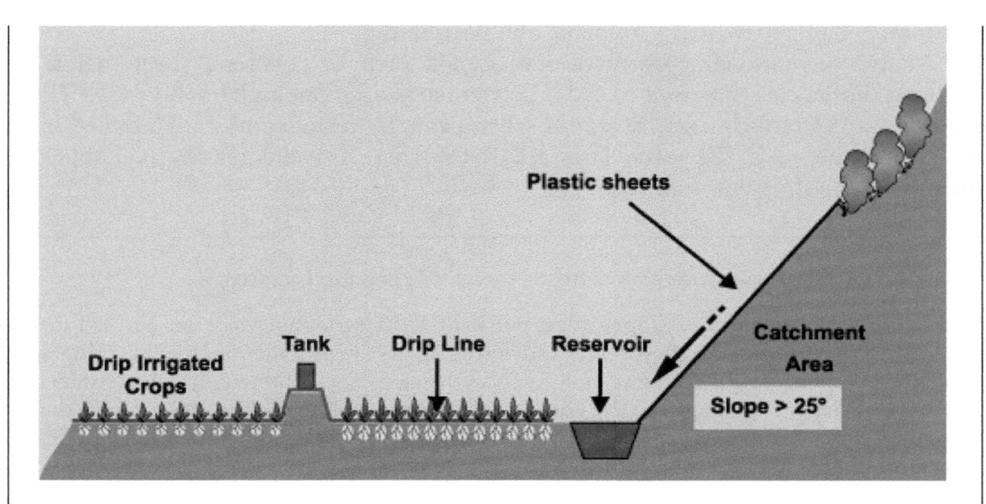

Figure B4.2.2 Collecting the runoff allows its utilization for drip irrigation of horticultural crops (Prinz, 2002a).

The plastic sheets are rolled up after the end of the rainy season and stored in the house until the start of the next rainy season. This protects them from damage by animals and sunlight.

Source: Prinz (2002a).

Another study in China evaluated runoff characteristics of six surface treatments relative to rainfall amount and intensity and antecedent rainfall during naturally occurring rainfall events in the semi-arid loess regions of northwest China. The surface treatments included two basic types, i.e. earthen (natural loess slope and cleared loess slope) and barrier-type (concrete, asphalt–fiberglass, plastic film, and gravel-covered plastic film). The results indicated that runoff and runoff efficiency of the earthen surface treatments were closely related to rain intensity, while runoff from the asphalt–fiberglass, plastic film, gravel-covered plastic film, and concrete surface treatments was more governed by the amount of rainfall. Asphalt–fiberglass had the highest average annual runoff efficiency of 74–81%, followed in decreasing order by the plastic film (57–76%), gravel-covered plastic film (56–77%), concrete (46–69%), cleared loess slope (12–13%), and natural loess slope (9–11%) (Figure 4.8). Antecedent rainfall had an obvious effect on runoff yield for the cleared loess slope, natural loess slope, and concrete. The threshold rainfall was 8.5, 8.0, and 1.5 mm for the natural loess slope, cleared loess slope, and concrete treatment, respectively, without antecedent rainfall effects, and 6.0, 5.0, and 1.2 mm, respectively, with antecedent rainfall effects. Due to the impermeable surface, antecedent rainfall had little effect on the runoff yield for the asphalt–fiberglass, plastic film, and gravel-covered plastic film treatments, which had threshold rainfall of 0.1, 0.2, and 0.9 mm, respectively (Xiao-Yan *et al.*, 2004).

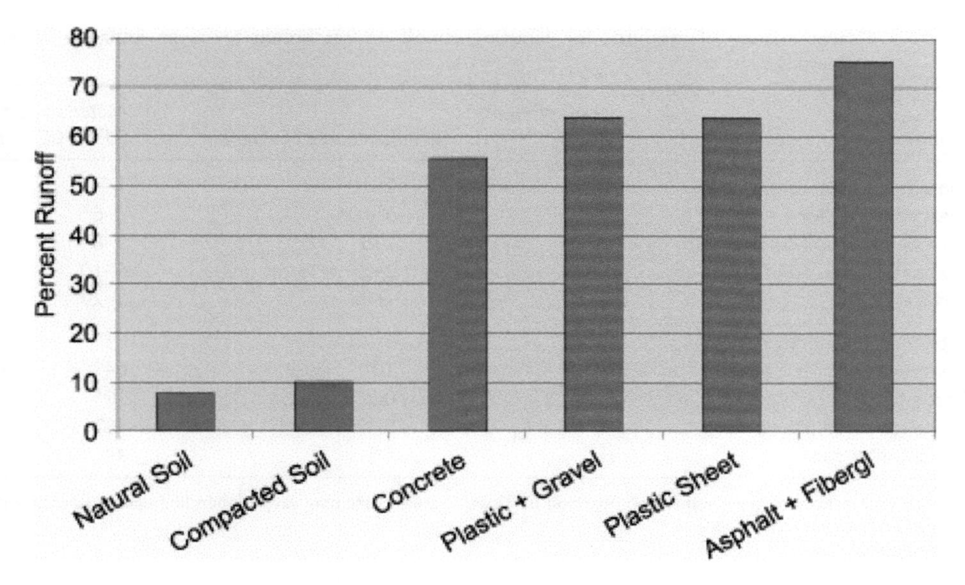

Figure 4.8 Annual runoff (%) from various artificial catchment materials in the semi-arid loess region of China (Xiao-Yan *et al.*, 2004).

4.3 ADVANTAGES AND DISADVANTAGES OF RUNOFF-INDUCEMENT METHODS

Many runoff inducement methods are site and area specific. The best method to use for a specific area depends upon the catchment characteristics, topography, rock type, and soil conditions. The socioeconomic conditions of the farmers are also an important consideration, as is government policy. Myers & Frasier (1975) list some of the desirable characteristics of catchment-area treatments to be used for runoff inducement:

- The resulting surface of the treated area should be relatively smooth and impermeable to water.
- The treated catchment area should have a high resistance to weathering damage (hot and cold) and resistance to deterioration from internal chemical and physical properties.
- The treatment should be able to resist damage by hail, intense rainfall, wind, occasional animal traffic, moderate flow of water, plant growth, insects, birds and burrowing animals.
- The treatment should be inexpensive on an annual cost basis, and should permit minimum site preparation and construction costs.
- Operation and maintenance should be simple and inexpensive, and the lifespan of the treatment should be as long as possible.
- Runoff water collected from the treated area must be nontoxic to plants and should not be harmful to human health.

Table 4.1 Characteristics of methods for inducing runoff on catchment areas to enhance water harvesting.

Treatment	*Runoff efficiency (%)*	*Estimated life (years)*	*Materials cost (US$/100 m²)*
Clearing of catchment	10–15	2–10	1–4
Cleared loess slope	12*	5–10*	3–6*
Smoothing of soil surface	15–20	2–5	3–8
Compacting the soil surface	20–30	2–3	6–10
Surface sealing	30–80	3–6	5–20
Impermeable coverings			
Concrete	69*	10*	64*
Plastic film	76*	0.5**	19*
Gravel-covered plastic film	77*	0.8**	38*
Asphalt–fiberglass	81*	5*	44*

Material cost estimates are given for comparison purposes. Actual costs may vary depending upon site location and catchment characteristics.
* Source: Xiao-Yan *et al.* (2004).
** Plastic films not UV resistant.

Not all of these characteristics may be obtained with any one treatment. Table 4.1 lists design estimates of runoff efficiency, average expected life, and initial cost for material per unit area for some common catchment treatments for runoff inducement.

Some of the more-expensive methods have higher runoff efficiency (more than 90%) and longer life (15–20 years) than the less-expensive methods. More labor-intensive and cheaper methods usually have low runoff efficiency (10–20%). Figures 4.9 and 4.10 show the general trend of investment requirements and suitability of various runoff inducement techniques in relation to the type of water harvesting system. Selecting the most appropriate method still depends on an expert assessment of technical, cultural, socioeconomic, and political considerations.

4.4 FURTHER CONSIDERATIONS

Care is needed to minimize the side effects of runoff-inducement methods. Poorly designed and managed rainwater harvesting can lead to soil erosion, soil instability, and local flooding. However, data on rainfall intensity, variability and hydro-geology are lacking in many developing countries, which hampers selection of the appropriate method.

Soil erosion is a constant concern and can be controlled if the slope is short and not too steep. Slope of the drainage area affects the quantity and quality of runoff. In long-slope systems the most efficient water harvest is from a small, gently-sloping catchment with good soil conditions (or from steep catchments with rocky surface).

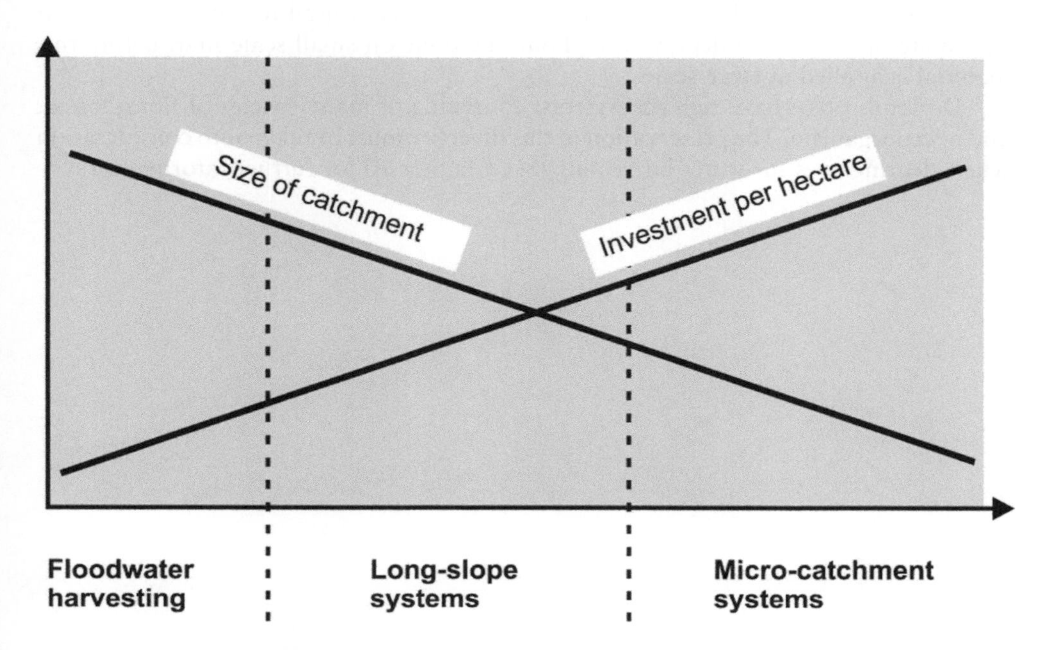

Figure 4.9 Relationship between catchment size and investment (labor and/or capital) for various runoff-inducing practices.

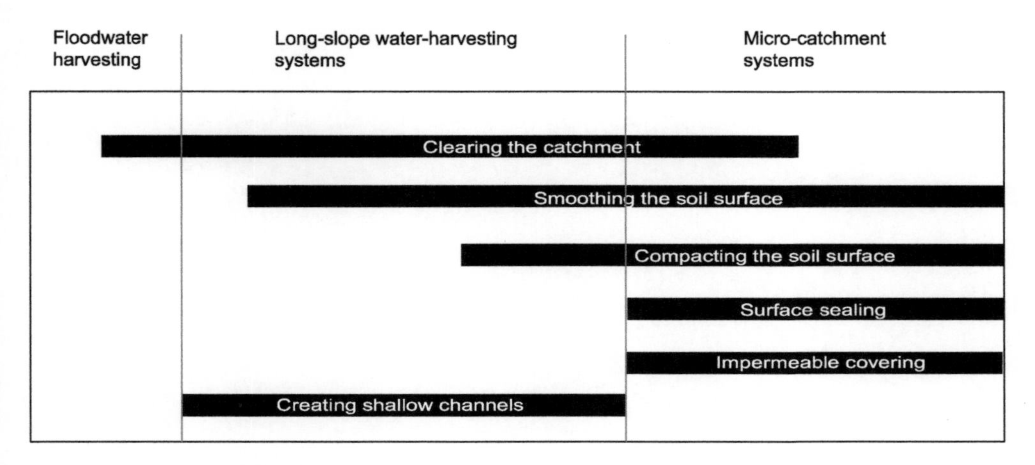

Figure 4.10 Suitability of runoff-inducement techniques to various water harvesting systems.

A rainwater harvesting catchment must withstand weathering and occasional traffic. However, most soil treatments have a limited lifespan and must be maintained and renewed periodically. They also require occasional maintenance because of cracking caused by unstable soil, oxidation, and weathering; plants growing up through the ground cover or the treated soil; and penetration by grazing animals.

Runoff water may be contaminated by the materials used to enhance runoff. If new materials are to be deployed, it should be done on small scale first, before this material is applied at large scale.

Drylands often have rich ecosystems, consisting of many species of flora, fauna, and microorganism. The preservation of this diversity must be taken into consideration when clearing sites for water harvesting (See Chapter 10 for further information).

Chapter 5

Identification of areas suitable for water harvesting

5.1 INTRODUCTION

Selection of appropriate sites and suitable methods under the prevailing conditions are among the most important prerequisites for successful water harvesting systems, particularly for the macro-catchment type which is characterized by large areas (Van Steenbergen *et al.*, 2011). Development of a methodology for identifying potential sites for RWH is an important step towards identifying areas suitable for certain techniques of water harvesting.

5.2 PARAMETERS FOR IDENTIFYING SUITABLE AREAS

Although water harvesting seems to be very simple, it cannot be implemented in all dryland areas. Parameters to be considered in identifying areas suitable for water-harvesting include:

- the prevailing climatic characteristics of the region, especially the rainfall;
- hydrology and alternative water resources;
- the topography of the region;
- the type of vegetation and agricultural production/ forestry activities;
- the soil type(s) of the region, including soil depth and soil fertility;
- socioeconomic conditions of the community;
- national laws (e.g. on water rights) and regulations, and
- infrastructural facilities available or planned for the area. (See Figure 5.1).

5.2.1 Rainfall characteristics

The availability of rainfall data collected over many years is crucial for the determination of the rainfall–runoff potential of a given region. This is particularly true in arid and semi-arid regions, where rainfall varies considerably from year to year. However, average rainfall can still be used in areas with insufficient rainfall data. Future changes due to Global Climate Change have to be anticipated.

Rainfall can be measured on site using non-recording rainfall gauges to record single rainfall events or the daily total rainfall in the project area. However, such data

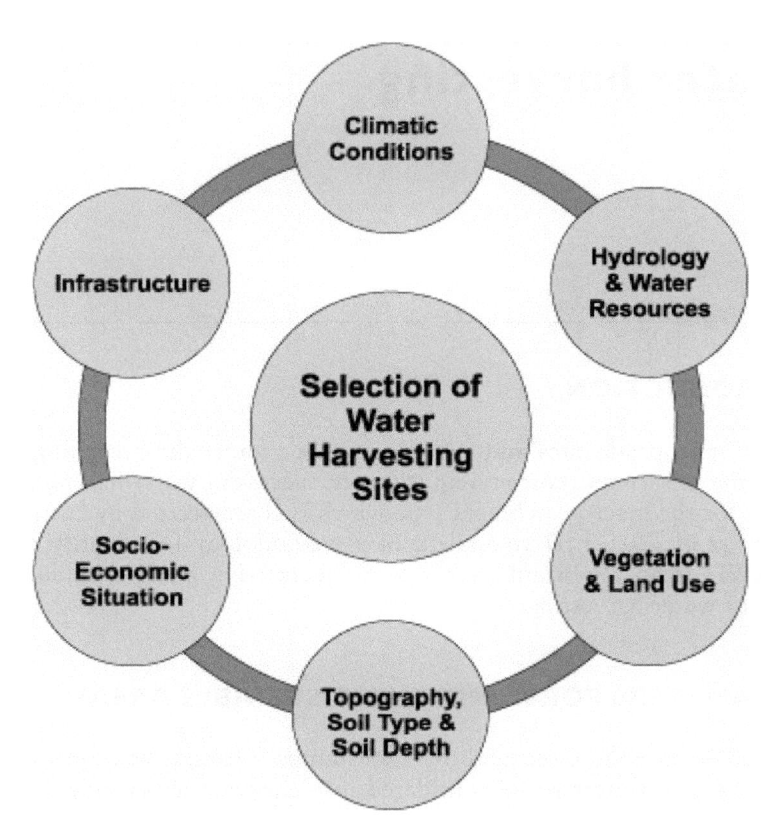

Figure 5.1 Main parameters to be considered in selecting areas suitable for water harvesting.

should be used with caution especially when extrapolating findings to adjacent areas: studies in a semi-arid region of the USA showed that 96% of the rainfall events at a station are representative of the rainfall of an area of only about 2.5 km^2 surrounding a rain gauge (Murphy *et al.*, 1977). Even the height of the rain gauge from the ground affects the amount of rainfall measured. To avoid discrepancies, rain gauges should be placed at the same height throughout the project area (Doorenbos, 1976).

Threshold rain, i.e. the depth of rain that must fall before runoff starts, is used in some rainfall–runoff models as a start value for runoff. Sufficient allowance must be made for the heterogeneity of the rainfall in time and space. Threshold rain varies with rainfall intensity, but soil type, inclination, vegetation, soil cover and antecedent soil moisture are further important parameters. For the Sahel region, threshold rain values between 3 mm and 10 mm have been reported. In Israel, for example, threshold rain was recorded as 3–4 mm for a dry soil surface, approximately half that for wet surfaces, 1–2 mm for solid rock and 3–5 mm for stony surfaces (Tauer & Humborg, 1992). Similar values might be found in other areas with similar climate and soils.

The intensity of the rainfall is a good indication of which rainfall is likely to produce runoff, although determining the threshold intensity that triggers runoff is more difficult than ascertaining the threshold rain depth. Rainfall intensity should also be determined as it is required for rainfall–runoff models. Recording rain gauges can be used for its determination.

Rainfall duration can also be determined reliably using a recording rain gauge. This is also an important factor because it is often related to peak discharge in simulation models.

Once these data have been acquired, the most important rainfall parameters to be determined are:

- the relationship between the storm intensity and its duration; and
- the number of storms per year, including their mean standard deviation and probability distribution.

These parameters will then be used to calculate the volume of water available for cropping, possibly by generating synthetic rain events for deterministic calculations of runoff quantities.

The temperature regime, air humidity and wind conditions during the cropping period are further climatic factors which have to be taken into consideration when selecting a certain area for water harvesting. Under Mediterranean climatic conditions or at higher altitudes the risk of frost occurrence during the rainy season might render affected areas unsuitable for water harvesting.

5.2.2 Hydrology and water resources

The hydrological processes relevant to water harvesting practices are those involved in the production, flow, and storage of runoff from rainfall within a particular catchment area. The intricacies of this phenomenon cannot be explained in detail here, but an overview is presented.

Rain falling on a particular catchment area can be divided into two major components: the effective rainfall for water harvesting (direct runoff) and losses. The sources of loss are:

- evaporation from the ground;
- water infiltration in the catchment;
- depression storage in the catchment; and
- water intercepted by leaves of plants.

In arid and semi-arid areas, extreme fluctuations in both annual rainfall and its distribution within the rainy season are a major constraint to agricultural production. In most cases, rainfall shows no regular patterns; wet periods are often followed by marked dry periods. Modeling the rainfall–runoff process in the hydrological analysis of an area is very complicated and the model designer must choose the most appropriate model from the existing model types or develop one to suit the area in question. The lack of meteorological data, of suitable topographical maps etc. often creates complications, limiting strongly the usefulness of models.

The availability of sufficient runoff that can be stored in the target area to meet the water requirement of the selected crops during the dry periods between rain events is a good indication of the suitability of the area for water harvesting.

Another factor to be taken into consideration is the availability of other water sources, e.g. of near-surface groundwater in *wadi* beds, of renewable or fossil groundwater from deeper aquifers. These water sources can either substitute runoff water during drought seasons or extend the cropping period beyond the rainy season.

5.2.3 Vegetation and land use

Vegetation strongly affects surface runoff. An increase in the vegetative density results in a corresponding increase in losses to interception, retention, and infiltration, which consequently decreases the volume of runoff. A typical example of what might be obtainable in many arid and semi-arid region is depicted in Figure 5.2. The density of vegetation on a given area can be determined in a variety of ways, but remote sensing is useful if the project area is large and funds are available. Remote sensing uses the different reflectance of soil and vegetation as an indicator of the density of the vegetation (Tauer & Humborg, 1992).

Land use affects in various ways the suitability of land for water harvesting. Introducing micro-catchment harvesting in areas already under cropping is much easier than transferring farmers into potentially suitable areas. On the other hand, farming activities in catchment areas reduce significantly the runoff yield, as plowed fields show high infiltration rates. Grazing reduces the vegetation cover; this will result in higher runoff volumes of catchment areas. Overgrazing, however, entails in most cases a raised soil erosion risk, with negative impacts on the water harvesting potential of the region.

5.2.4 Topography, soil type and soil depth

The suitability of an area for water harvesting depends strongly on its topography and soil characteristics, namely:

- the slope of a terrain, which is a decisive factor for any type of water harvesting;
- surface structure, which influences the rainfall–runoff process;
- infiltration and percolation rates, which determine the movement of water into the soil and within the soil matrix; and
- soil depth, which, together with the soil texture, determines the quantity of water that can be stored in the soil.

The topography has a strong impact on infiltration volume and runoff yield. Microcatchment systems are more appropriate for gentle slopes, whereas macro-catchment techniques can only be established in terrain having significant slopes (Tauer & Humborg, 1992). Further information is given in Chapter three.

The infiltration rate is the amount of water entering the soil, through its surface, over a given time. Infiltrometers and/or rainfall simulators can be used to determine the infiltration behavior of any soil. The main soil parameters affecting infiltration rate are texture, structure, and depth (Figure 5.3; see also chapter 6.2.1). Vegetation and soil fauna also affect infiltration rate. Dense vegetation protects the soil surface and increases water retention, increasing infiltration rate. A well-developed root system also increases infiltration rate. If the soil is bare of vegetation, raindrops hit the soil surface directly, sealing the surface, which hinders infiltration.

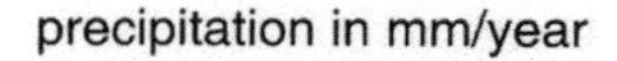

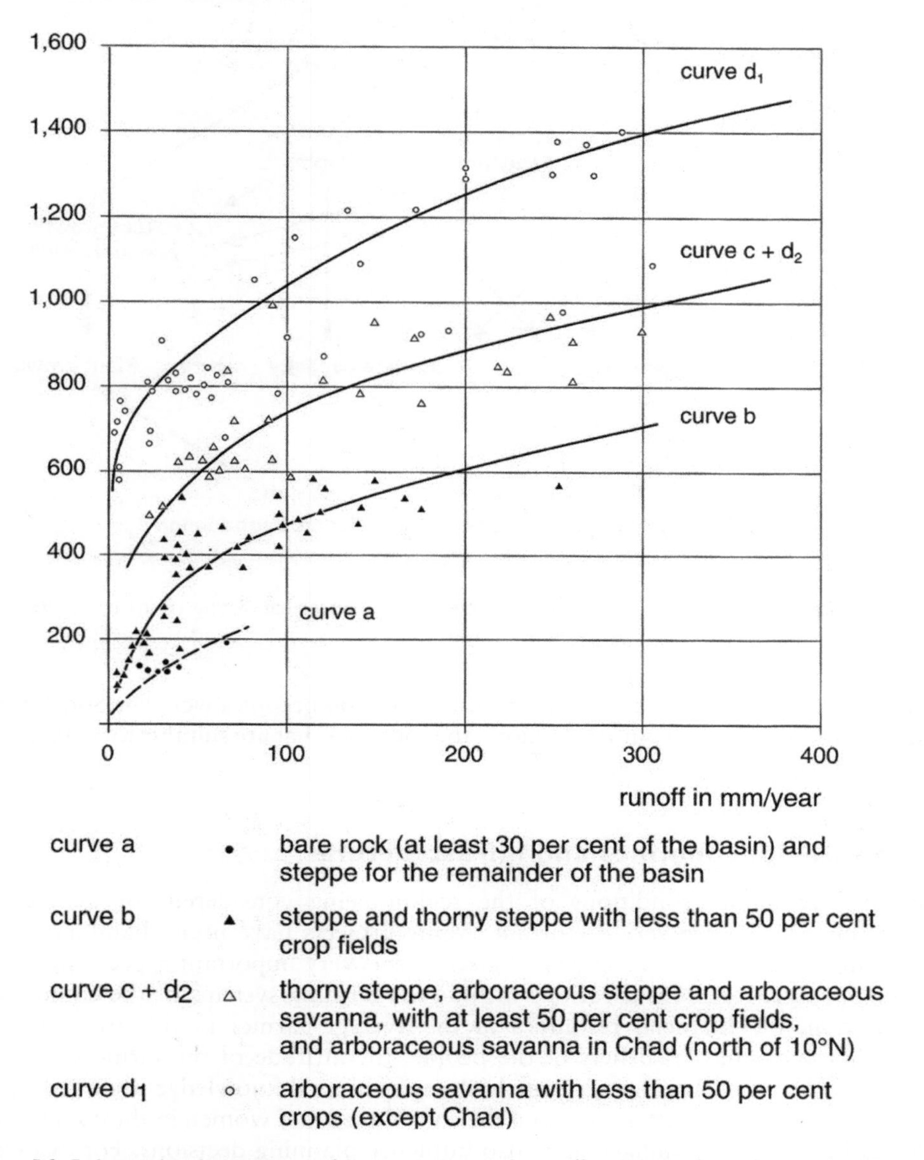

Figure 5.2 Relationship between rainfall, vegetation and runoff in West Africa. (Tauer & Humborg, 1992; p. 76, based on Davy *et al.*, 1976).

Initial infiltration rates are higher on dry soils than on wet soils. As rain falls the infiltration rate declines rapidly because pores near the surface fill quickly and the hydraulic gradient, which is the driving force for the infiltration process, drops rapidly. In addition, soil surface sealing may occur, reducing infiltration. The cracks

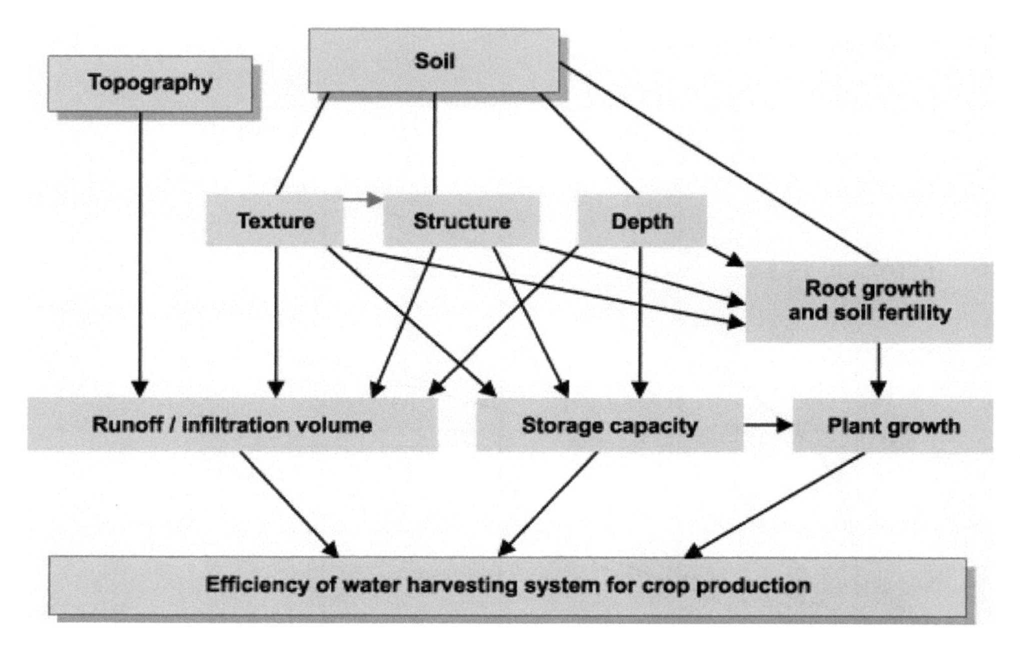

Figure 5.3 Soil properties strongly influence the efficiency of water harvesting. (Prinz *et al.*, 1996a).

that frequently occur in clay-rich soils close as the soil becomes wet. The Soil Texture Classification Triangle (Figure 5.4) shows the soil types that are suitable for catchments and for cropping (Prinz *et al.*, 1999).

5.2.5 Socioeconomics and infrastructure

The socioeconomic conditions of the region being considered for any water-harvesting scheme are very important. Many projects have been abandoned soon after implementation as a result of neglect of this very important aspect during the planning stage. Key considerations include the farming systems of the community in question, the financial resources of the average farmer in the area, cultural behaviors and religious beliefs of the people, the attitude of the farmers towards the introduction of new farming methods, the farmers' knowledge about irrigated agriculture, land property rights, and the role of men and women in the community. The mobility of the populace may also influence planning decisions. For example, in West Africa it was assumed that settled farmers would walk a maximum of 6 km from their homes to a water harvesting system, whereas seminomadic people would simply move to wherever they can find food and water for their animals irrespective of the distance.

As in any development projects, existing infrastructures or future plans that will be developed in the future on the same area have to be taken into account when planning a water harvesting scheme.

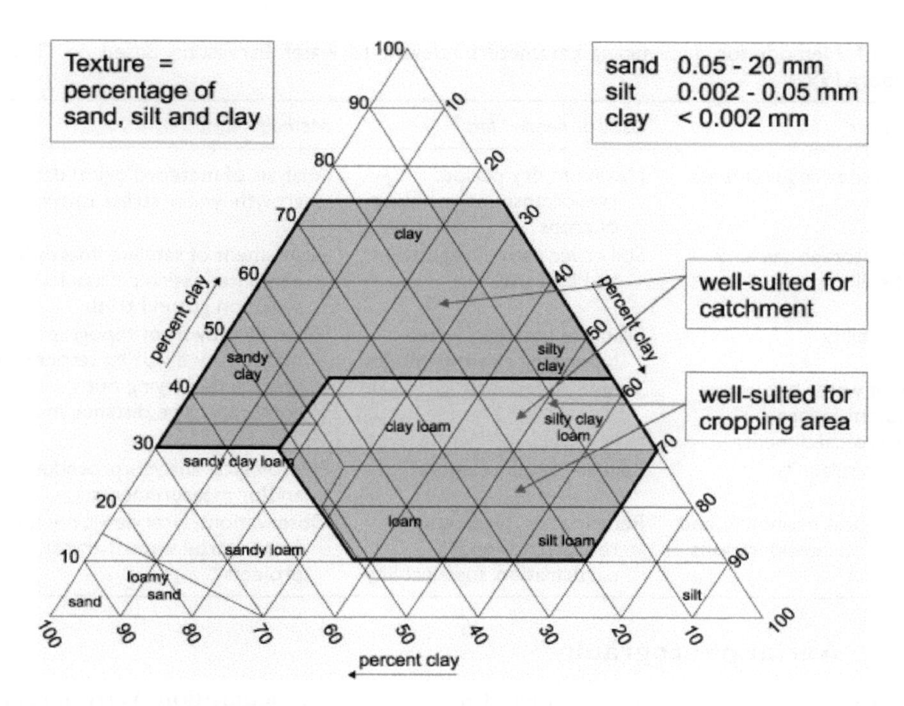

Figure 5.4 Soil types suitable for catchment and cropping areas in water harvesting systems. (Prinz *et al.*, 1996a).

5.3 METHODS OF DATA ACQUISITION

5.3.1 Overview

A number of basic data are required in any water harvesting project (Table 5.1). The choice of method used to acquire this data depends not only on technical and financial considerations, but may also be constrained by national security and political issues.

5.3.2 Ground truthing

Field visits to the area where a water harvesting project is to be executed are always necessary. For reliable results, specialists versed in hydrology, prevailing vegetative condition of the region, and possibly the agricultural practices of the local population will be needed. Ideally, these should be local experts. Some parameters cannot be directly ascertained from maps, aerial photographs, or even satellite images, but require an inventory of the terrain during field visits. Maps and ground truthing are adequate sources of information if the project will be executed in a small area but will be too time-consuming and expensive for larger areas or at regional levels.

Table 5.1 Methods for determining parameters relevant to water harvesting. Based on (Tauer & Humborg, 1992)

Parameter	*Used or needed for:*	*Method*
Crop water requirements	Maximum dry period, evapotranspiration values of crops	Analysis of meteorological data, plant growth–water stress relations
Water storage capacity of soil	Soil cover, natural vegetation, and land use	Assessment of satellite images by computer-assisted classification based on ground truth
Accessibility	Distance between water-harvesting site and villages	Taken directly from topographic maps or by digitizing settlement areas and carrying out comprehensive distance model
Type of water harvesting system (micro-/macro-catchment)	Terrain slope	
Water availability	Rainfall–runoff relationships	Hydrological analysis/procedures and/or measurements
Sociological, economic, and political considerations	Beneficiaries' preferences, resources support, participation, sustainability	Observations, interviews, outcome from nearby water harvesting projects

5.3.3 Aerial photography

Aerial surveying is a proven instrument for extensive data acquisition. Vertical surveys with stereoscopic overlap can be made using large cameras.

Aerial surveying depends on the regional or national availability of survey planes, and is cost-effective only for large-scale projects. It may be appropriate for a water-harvesting project on a regional scale.

5.3.4 Satellite and remote-sensing technology

Satellite and remote-sensing technologies coupled with geographical information systems (GIS; see chapter 5.4.3 for more information) are the most powerful and reasonably cost-efficient tools used in assessing the potential for water harvesting.

The term 'remote sensing' is used to describe all the procedures employed in recording information from high above the Earth's surface. This can be done from an airplane or satellite. Remote sensing technology can not only be used for preliminary gathering of information, but also to continuously monitor and update data at regular intervals.

Various types of information are available from a variety of remote-sensing platforms (Table 5.2).

The principal steps in using remotely sensed data to identify areas suited to water harvesting include:

- definition of data needs, e.g. land use, geology, pedology, hydrology, etc.;
- data collection using remote sensing and other techniques;

Table 5.2 Information for water harvesting planning obtainable from remote sensing systems.

Parameter	*Satellite type*	*Type of information*	*How to obtain*
Topography	Aerial photo, LIDAR, IfSAR	Raster data (DEM)	Internet sites, local mapping agencies
Inclination/ slope	Aerial photo, LIDAR, IfSAR	Raster data (percent, degree)	Derive from DEM
Elevation	Aerial photo, LIDAR, IfSAR	Raster data (meters above mean sea level)	Derive from DEM
Surface roughness	Microwave	Root mean square average	Microwave remote sensing
Soil type	Landsat TM, SPOT, ASTER, others	Polygons of soil mapping units	Interpretation and ground truthing
Soil depth	Ground penetrating radar	Raster (cm)	Interpretation and ground truthing
Soil moisture	Radar remote sensing	Raster (percent)	Interpretation and ground truthing
Land cover/ land use	Landsat TM, SPOT, ASTER, others	Polygons (classes)	Visual interpretation, image classification, and ground truthing
Type of vegetation	SPOT, ASTER, Hotbird, others	Polygons (type)	Visual interpretation, image classification, and ground truthing
Infrastructure	Aerial photo, Landsat TM, SPOT, ASTER, others	Vector data (points, lines, and polygons)	Visual interpretation and ground truthing (local mapping agencies)
Water bodies	Aerial photo, Landsat TM, SPOT, ASTER, others	Polygons	Visual interpretation and image classification

- data analysis, e.g. measurement, classification, and estimation analysis;
- verification of the analysis results; and
- presenting the results in a suitable format, such as maps, computer data files, written reports with diagrams, tables, maps, etc.

Water, forest, pasture, and other features reflect light from the sun differently and yield characteristic patterns in the relation between wavelengths and amount of reflected energy. These patterns can be recognized in the data registered by the satellite. Image classification is based on the assumption that areas with similar characteristic spectra have similar characteristics on the ground. There are two approaches to classification of the data that are distinguished primarily by their initial assumptions. In supervised classification, ground truth data from direct in-field observations are used to identify the initial parameters used in classification. Unsupervised classification uses no ground truth in its initial stages, but the final images and maps produced must almost always be verified in the field.

In remote sensing cartography, the acquired information is first classified in problem-oriented categories, and is then mapped in accordance with standard cartographical rules. Compared to approaches using aerial photography and ground truth, a less effort is required to process remotely sensed data because certain stages of the analysis can be assessed on the monitor to elaborate certain statistical evaluations. Since the data gained through this system is in digital form, it can be translated to adjacent scenes with the consideration of the existing illumination differences without the need to carry out field investigations. Since the remotely sensed data are in digital form, it can be further processed and even linked to other compatible data sets (Oberle, 2004; Tauer & Humborg, 1992).

Use of Google Earth imagery

Google Earth images are now available worldwide and offer good opportunities for water harvesting planning. Box 5.1 presents an example of the application of Google Earth images. The images are available free of charge, but a commercial version that offers more features is also available. The drawbacks to Google Earth images are that they are normally several years old and changes are not recorded. Nevertheless, Google Earth images can serve as basis for land-use planning.

Box 5.1 Application of Google Earth technology in planning a water harvesting project.

In 2010, ICARDA in cooperation with the Agricultural Research Council of Libya carried out a study to explore the potential of water harvesting in northern Libya (Prinz, 2010). To identify locations (mainly farms) suitable for water harvesting research and demonstration activities, a number of GIS maps were developed by ICARDA, integrating available knowledge on topography, soils, precipitation, slopes, infrastructure, etc. These maps were used to identify areas of high potential for certain water harvesting techniques.

Ground truthing included determination of slopes, soil type and soil depth, farm sizes, accessibility, and acceptability of proposed actions to farmers, has led to the identification of 16 farms, one research station, and four other sites suitable for activities such as reforestation, dam construction, etc.

As the time available did not allow a topographic survey of all locations selected, Google Earth images were the only readily available source of information for the detailed planning of research and demonstration measures.

The quality of 'raw images' from Google Earth varies considerably according to the satellite system used and date of exposure.

A high-quality image from Google Earth (Figure B5.1.1) allows identification of topographic features, such as bunds and ditches, infrastructure (e.g. roads and buildings), land use, such as irrigated and rainfed areas, and even some types of vegetation, such as crops, trees, and bushes.

Such images can serve as the basis for land-use planning, and, as in Libya, for detailed planning of spatial distribution of various water harvesting interventions (Figure B5.1.2).

Figure B5.1.1 Topographic features, infrastructure, and even vegetation show up in high-quality images from Google Earth. (Prinz, 2010). (*See color plate, page 250*).

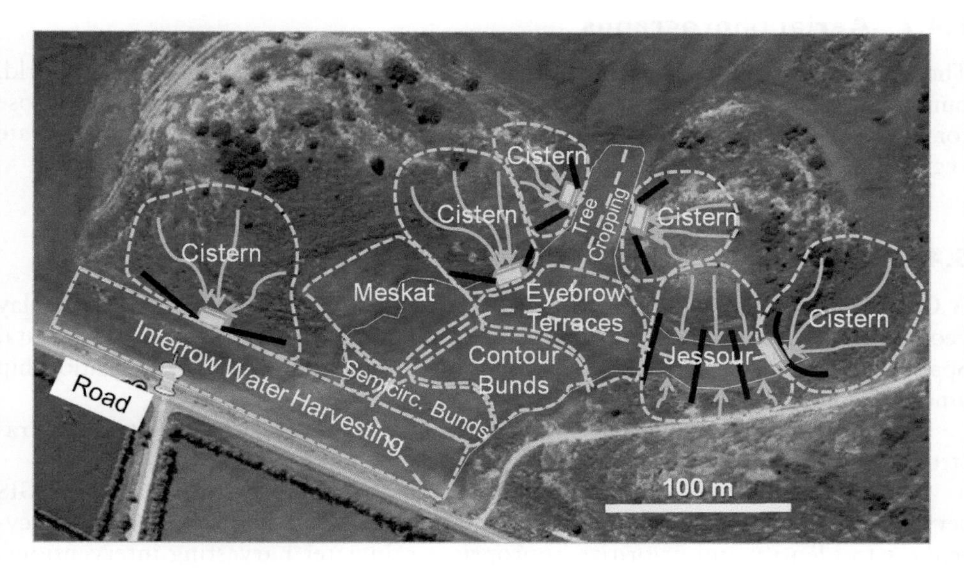

Figure B5.1.2 A Google Earth image used to plan water harvesting interventions in Libya. (Prinz, 2010). (*See color plate, page 250*).

5.4 TOOLS

5.4.1 Maps

Maps may still be the only means of acquiring data in some countries, aside from Google Earth images. Two types of maps have commonly been used in gathering land-related information: topographic and thematic.

5.4.1.1 Topographic maps

A topographic map represents the features of an area in an analogue form. This type of map can be found in many regions of the world. They can be digitized and incorporated into a GIS database.

5.4.1.2 Thematic maps

Thematic maps present specific types of information, e.g. soils, rainfall or temperature isohyets, vegetation types, etc. Thematic maps present source information in classes. It should be noted that a degree of inaccuracy exists in the way classes are defined. For instance, a continuous phenomenon such as soil or vegetation type is mapped as homogeneous map units with sharp boundaries, whereas the actual circumstances on the ground vary within each map unit; this may affect project results significantly.

5.4.2 Aerial photographs

There are archives of black and white aerial photographs in many parts of the world, but their usefulness depends on the age and scale of the images and the specific purpose for which they were taken. Color-infrared photographs can be used to differentiate vegetation types (Mati *et al.*, 2006).

5.4.3 Geographic information systems

A GIS is a computer-based system used to capture, store, edit, manage, and display geographically referenced information, including spatial and descriptive data. Spatial data deal with the location and shape of various features and the relationship among them.

Such features as topography, water resources, soil types, land-use types, infrastructure, and administrative boundaries can be combined in a GIS.

Descriptive data describe the characteristics of these features. Thus, a GIS serves as a tool for representing the real world. GIS can be used to help policy-makers to identify and prioritize appropriate rainwater harvesting interventions. Examples of the use of GIS in water harvesting projects are given in Boxes 5.2, 5.3, and 5.4.

Box 5.2 Determining Africa's water harvesting potential.

The World Agroforestry Centre (ICRAF) and UNEP have created thematic GIS databases using baseline and hydro-physical data on rainfall, topography, soils, human settlements, and land use. These data were combined with information on agroclimatic zones, soils, population densities, and hydrological data to create composite maps, displaying so-called development domains that corresponded with different levels of suitability for certain types of water harvesting technologies such as:

- rooftop and surface runoff with storage in ponds or pans;
- flood-flow from water courses with storage in subsurface dams or sand storage dams; and
- *in situ* soil-water storage systems.

In total, 98 thematic maps were developed, 29 with continental coverage (see Figure B5.2.1) and 69 of 11 selected countries: Botswana, Ethiopia, Kenya, Malawi, Mozambique, Rwanda, Somalia, Tanzania, Uganda, Zambia and Zimbabwe.

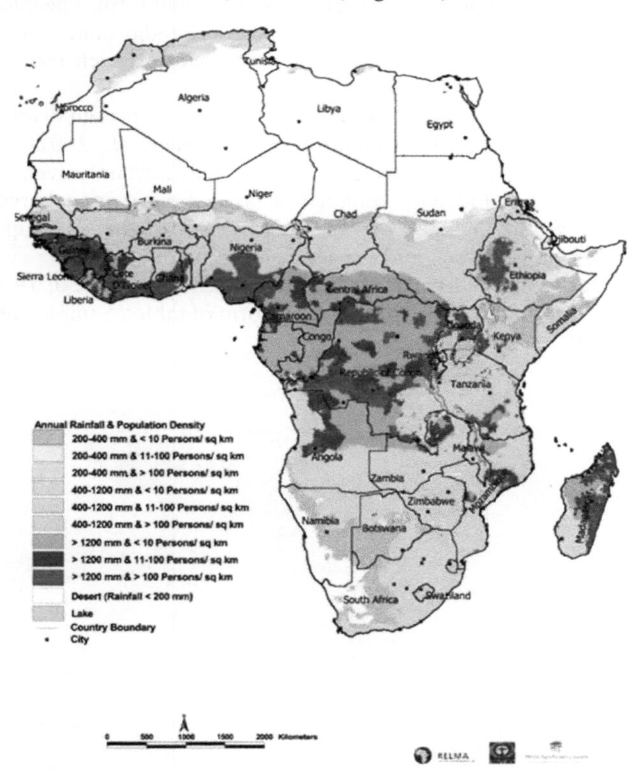

Figure B5.2.1 Map showing potential for runoff-water harvesting across Africa. (Mati *et al.*, 2006). *(See color plate, page 251)*.

In addition, an in-depth spatial analysis of water harvesting potential was carried out for Addis Ababa (Ethiopia), Kampala (Uganda), Lusaka (Zambia), and Nairobi (Kenya).

This GIS database – the first of its kind for Africa – will be used to help policymakers to identify and prioritize appropriate rainwater harvesting interventions in Africa.

Sources: http://www.unep.org/pdf/RWH_in_Africa-final.pdf
http://www.treesofchange.org/downloads/publications/PDFs/MN15297.PDF

Box 5.3 Using GIS to plan water harvesting interventions in Syria.

Potential for water harvesting in Syria was assessed by using a GIS to match simple biophysical information, systematically available at country level, to the broad requirements of the specified water harvesting systems (De Pauw *et al.*, 2008). The systems evaluated include 13 micro-catchment forms, based on combinations of seven techniques and three crop groups, and one generalized macro-catchment form. The micro-catchment techniques were: contour ridges, semicircular and trapezoidal bunds, small pits, small runoff basins, runoff strips, inter-row systems and contour-bench terraces. The three crop groups were rangeland, field crops, and fruit trees.

The environmental criteria for suitability were based on expert guidelines for selecting water harvesting techniques in drier environments (Oweis *et al.*, 2001). They included precipitation, slope, soil depth, texture, and salinity, land use/land cover, and geological substratum. The dataset included interpolated surfaces of mean annual precipitation, the Shuttle Radar Topography Mission digital elevation model, the soil map of Syria, the land use/land cover map of Syria, and the geological map of Syria.

The results of the suitability assessments are presented as a set of 14 maps and are summarized at provincial and district level in the form of tables. Sample output maps are shown in Figures B5.3.1 and B5.3.2.

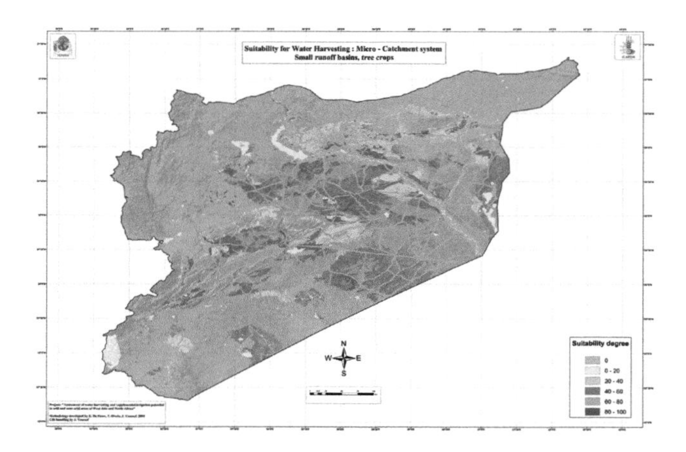

Figure B5.3.1 Map showing suitability for water harvesting using contour ridges to grow range shrubs. (*See color plate, page 252*).

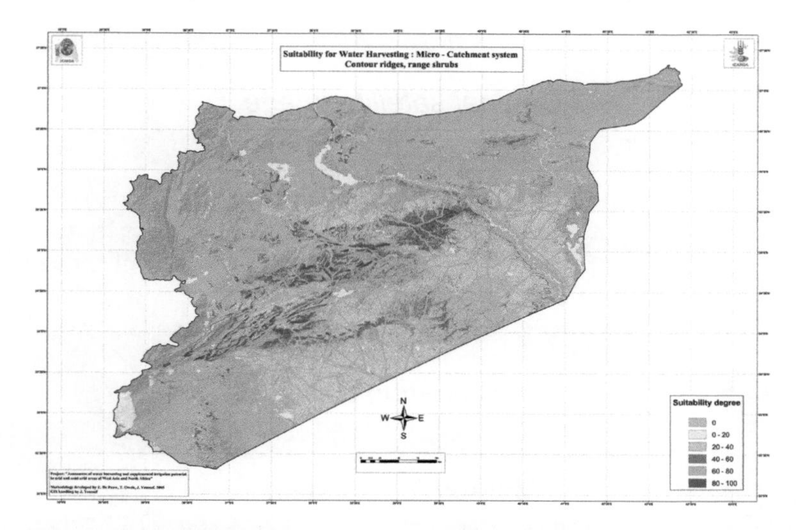

Figure B5.3.2 Map showing suitability for water harvesting using small runoff basins to grow tree crops. (*See color plate, page 252*).

The results need to be validated by comparing the results predicted by the model with an assessment of actual conditions in sample locations.

Source: De Pauw *et al.* (2008).

Creation of GIS for area identification

The type of information needed may differ depending on the type of water harvesting to be implemented. However, four types of basic data will be required for a GIS database:

- Existing maps of the area
- Remote-sensing data
- Existing databases, e.g. climatic data
- Data from field visits and measurements.

All the above data types have to be identified, selected, and classified. They are then stored in a computer to form a GIS database containing geometrical and attribute information (Figure 5.5).

Creating a GIS database is both costly and time-consuming, but once the database is created it can be analyzed using a variety of GIS manipulation tools. Some of the commonly employed manipulation tools are:

- Network analysis
- Digital terrain analysis
- Cartographic analysis, overlay, and intersection

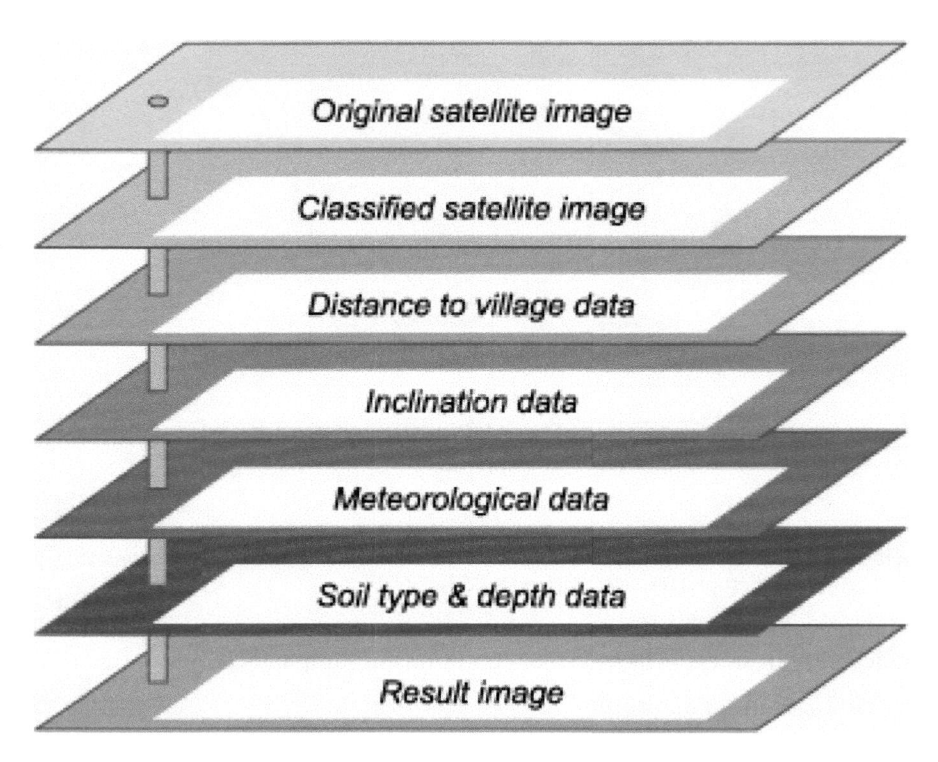

Figure 5.5 The principle of preparing a GIS based on satellite images and digitized topographic maps. (Tauer & Humborg, 1992, redrawn and altered).

- Model applications connected to the GIS database
- Statistical analysis and classification
- Digital image processing.

The results of the analysis are finally presented in the form of graphs, maps, plots, statistics, and scripts. The decisive factor in terms of determining the most-suitable water harvesting technique for a particular site is the slope of the terrain. An inclination model may be created from a digital terrain model (DTM), but creating a DTM requires considerable effort and is time-consuming and costly if the area of interest is large. The cost is also influenced by the method used to produce the DTM. The most important methods used in obtaining a DTM are (in ascending order of cost):

- digitization of topographic maps;
- photometric assessment of satellite images;
- photometric assessment of aerial photographs; and
- *in situ* topographic survey.

The most accurate, and expensive, method is a tographic survey using geodetic instruments. However, digitization of topographic maps is recommended for many developing nations to reduce costs. The terrain data from available maps may not be good enough for a hydrologic model, but can be used with acceptable accuracy in predicting the type of water harvesting to be established in a given area.

5.5 DECISION TREES

Decision Trees are excellent tools for helping to choose between several courses of action. They provide a highly effective structure within which one can lay out options and investigate the possible outcomes when choosing those options. They also help to form a balanced picture of the risks and rewards associated with each possible course of action. A value or score is assigned to each possible outcome. Decision tree analysis requires an estimate of the probability or possibility of each outcome. If we use percentages, the total must come to 100%. If we use fractions, these must add up to 1. If we have data on past events one may be able to make rigorous estimates of the probabilities, otherwise best guess is made. After working out the value of the outcomes, and have assessed the probability of uncertainty, it is time to start calculating the values that will help us make our decision.

Decision trees or diagrams can help in determining the most appropriate site to choose for a particular water harvesting system. For example, a project in Mali, West Africa, used a decision diagram to determine the water harvesting potential of the area (Tauer & Humborg, 1992). Satellite images of the water storage capacity of the soil together with the inclinations and digitized topographical maps indicating distances between populated centers and potential water harvesting sites were superimposed on three levels through a GIS raster. If all types of water harvesting system are grouped into macro- and micro-catchment water harvesting systems, one can be able to find the most appropriate technique in a given area (Figure 5.6).

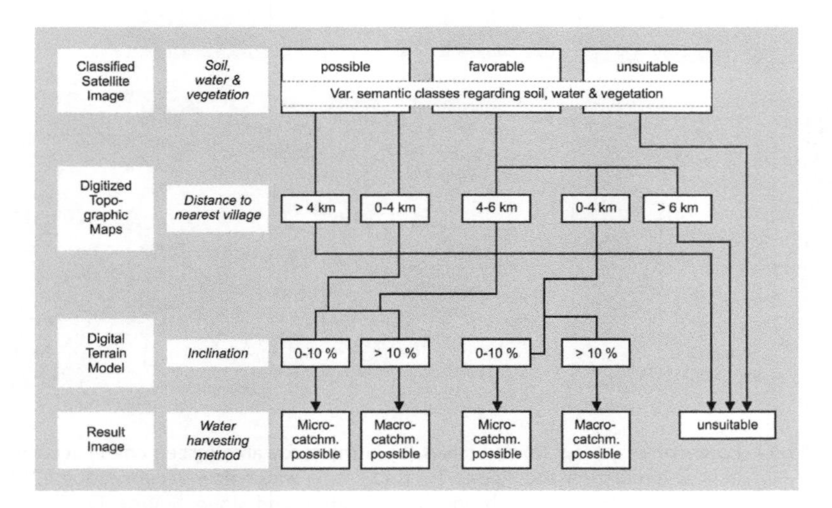

Figure 5.6 Decision tree used to identify potential water harvesting sites in the northern Kayes Region in Mali (Tauer & Humborg, 1992, redrawn).

Box 5.4 Deriving suitability for Water Harvesting structures in Tunisia.

In Tunisia, a study concerning the development of a methodology for identifying areas potentially favorable for water harvesting was carried out. Currently available data and knowledge together with modern tools such as image processing and geographic information systems to map potential for water harvesting were used. For reasons of practicality, the selected systems were limited to the *jessour* and *tabias* techniques which are widely used in the mountainous area of southern Tunisia, where annual rainfall is below 250 mm. *Jessour* are based on the use of a large area with an significant slope for runoff and the development of agricultural activities in the run-on areas on the alluvial deposits. *Tabias* are simply similar structures but used in the foothills and surrounding plains.

Considering that water harvesting is concerned with the management of surface water runoff, the procedure emphasized the importance of defining physical conditions favorable to implementing the selected structures. In order to gain eventually wide application in the region, the input data have been limited to the commonly available spatial and climatological data.

Potential area for water harvesting was obtained by combining suitability criteria of flow accumulation and slope. Results obtained for *jessour* and *tabias* are coherent with the data reported on the land use map based on field surveys. (Figure B5.4.1)

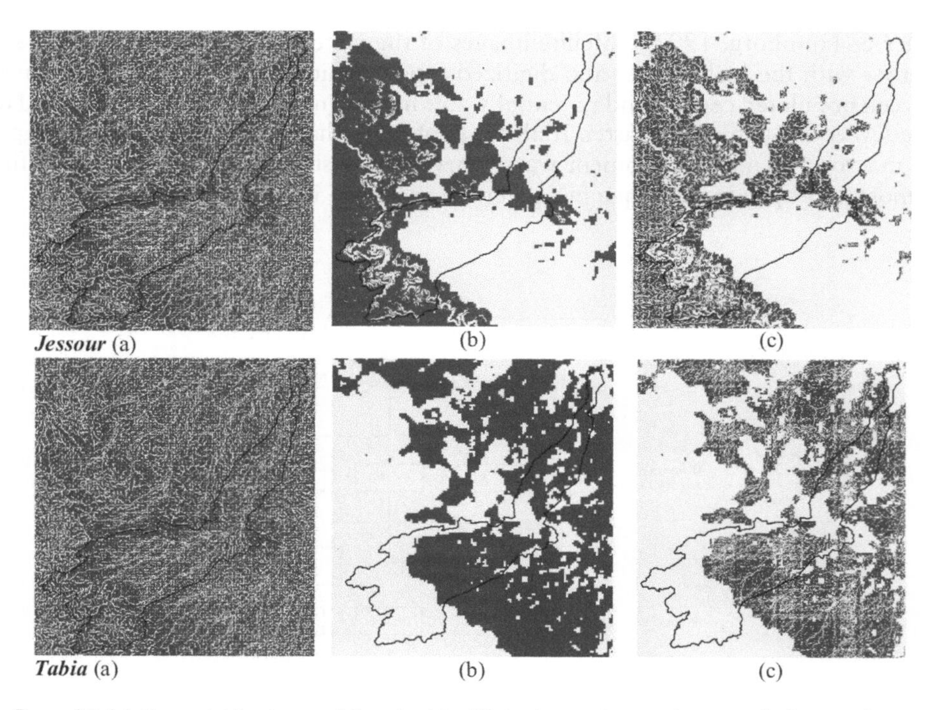

Figure B5.4.1 Potential for *jessour* (a) and *tabias* (b) in the study area (green color) according to flow accumulation and slopes. (1) Using only water flow accumulation, (2) Using only slope, (3) Using both flow accumulation and slope. Source: De Pauw *et al.* (2008). (*See color plate, page 253*).

Chapter 6

Planning and design of water harvesting systems

6.1 INTRODUCTION

Plant growth is influenced by a wide range of physical, climatic, and biological factors. In dry areas, however, water shortage may be the principal limiting factor on crop production. Some understanding of soil–water–plant relations is necessary for design and management of water harvesting interventions. This includes an understanding of the significance of water to plants and the way in which the plants obtain and lose water in the atmosphere–soil–plant system (Berliner, 2005; Barrow, 1999). Almost every process occurring in plants is affected by water, but the links are complex. The plant–water relationship varies with plant characteristics, stage of development, soil, and climatic conditions. In simple terms, water is absorbed by the plant roots and lost by transpiration. About 95% of water taken in by plants is transpired, while 5% or less used in plant growth. If water cannot be absorbed by the roots to compensate for transpiration loss, then a water deficit develops in the plants, and plants may wilt. Water shortage has a significant impact on plant growth and production, but the impact magnitude may vary depending upon plant type and stage of development.

6.2 SOIL–WATER–PLANT–CLIMATE RELATIONS

6.2.1 Soil

Soil is a three-dimensional body, occupying the upper most part of Earth's crust. Its properties differ from those of the underlying rock as a result of interactions between climate, living organisms, parent material, and relief over period of time. Soils are classified according to various taxonomic systems (e.g. FAO-UNESCO, US Soil Taxonomy, Russian Soil Taxonomy). They are usually grouped into taxonomic categories according to their properties, which reflect the nature and stage of their development. Soils belonging to one taxonomic category are homogeneous within the range of that category. The most important soil properties and conditions relevant to water harvesting and irrigation are texture, structure, depth, and infiltration. In regard to runoff generation the effect of soil crusting cannot be underestimated (Carmi & Berliner, 2008).

6.2.1.1 *Texture and structure*

The most important property influencing the soil characteristics is soil texture. This refers to particle size distribution and particle size groupings. Soil texture is the most permanent characteristic of the soil and influences a number of other soil properties, such as structure, consistency, moisture regime, permeability, infiltration rate, runoff rate, erodibility, workability, root penetration, and fertility.

Soil texture is determined by the percentages of sand, silt, and clay in the soil. Clayey soils (more than 30% clay) are well suited for catchments while loamy soils are best suited for the cropping area (see Figure 5.4).

Soil structure refers to the arrangement of primary soil particles (sand, silt, and clay) and organic substance and their formation into natural aggregates. Soil structure exerts a dominant influence on the soil, air, and moisture regime, on the soil's hydraulic conductivity, and consequently on the root growth and microbial activities that occur within the soil. It is, therefore, an important factor in soil productivity and soil development. Clay and organic matter serve as aggregating agents. The pore space that is formed within and between soil aggregates is taken up by air and moisture and allows the penetration and passage of plant roots. The wider voids that cannot hold water promote the circulation of air and the drainage of excess water.

The ratio of the approximate volumes of solid, soil moisture, and soil air for an ideal structure are 2:1:1, while in practice it may be in between 1.5:1:1 and 1:1:1. A structureless soil is often characterized by compaction, poor aeration, low hydraulic conductivity, and poor root penetration.

The structure of topsoil is particularly vulnerable to the harmful effects of certain external factors. Drought, heavy rainfall, mechanical mishandling, and certain fertilizers may all adversely affect the soil structure. For example:

- Drought may reduce the organic matter content of the soil and cause sandy and silty topsoil to become loose dust and clayey topsoil to become brick hard.
- Heavy rain may destroy soil aggregates through the impact of the raindrops, resulting in crust formation and/or runoff. This is one of the major causes of soil erosion. Protection of the soil surface by plant cover or mulch is necessary to prevent such erosion.
- Mechanical mishandling of an over-wet soil, e.g. trampling by livestock or mechanical soil tillage, will have a negative influence on soil structure.
- Fertilizers containing sodium or, to a lesser extent, ammonium may harm the structure of certain soils that contain little organic matter.

6.2.1.2 *Water-holding capacity and soil depth*

Soil holds capillary water within and between the soil particles and on their solid surfaces due to adhesion and surface-tension forces. Plants cannot withdraw or absorb all water existing in the soil. The water content of the soil at (and below) which the plant can no longer withdraw water is known as the wilting point (WP). The term 'field capacity' (FC) means the water content of a soil that has been thoroughly saturated with water then left to drain freely until drainage practically ceases. Although FC depends only on soil, WP is affected by two major interacting factors: the soil and the plant. The influence of the plant is less than that of the soil.

The water content of a soil is determined either in the laboratory or in the field. It is usually expressed as a percentage mass (or weight) of the oven-dry soil as follows:

$$P_m = (M_w/M_s) \times 100$$

where:

P_m is the percentage mass
M_w is the mass of water, and
M_s is the mass of the soil sample after oven drying at 105°C.

In agricultural practice, it is often necessary and more useful to express the water content as a percentage by volume of the undisturbed soil. This can be done by multiplying the water content on dry mass basis by the apparent specific gravity of the soil as follows:

$$P_v\ (\%) = P_m \times \text{apparent specific gravity}$$

where:

P_v is percentage volume of water.

If the mass density of water is taken as 1 g/cm^3, the apparent specific gravity is numerically identical to the mass bulk density of soil.

In irrigation, it is customary to express volume percentages (P_v) in depth of water per given depth of soil. Thus, a P_v of 1% (one unit volume of water per hundred volume units of soil) means one depth unit of water per hundred depth units of soil, e.g. 1 mm of water per 100 mm of soil depth. This is also equivalent to 0.1 mm of water per 1 cm of soil depth.

The capillary water of soil between FC, as an upper limit, and WP, as a lower limit, is known as available water, AW. Thus, AW is the difference between FC and WP. For example, if FC (by volume) = 30%, and WP (by volume) = 16%, then AW = 14%. This AW is equivalent to 14 mm of water in 100 mm depth of soil, or 140 mm of water in 1 m depth of soil.

These forms of expressing available water are very useful and common in irrigation and farm water management. Available water of soil is also referred to as water-holding capacity, water-storage capacity, and water-retention capacity. The following text uses the term water-holding capacity (WHC). Table 6.1 presents typical WHC values for various soil textures.

The total depth of water available to the crop, TAW, depends on WHC and depth of soil from which roots can take water (effective root zone depth). TAW can be expressed as:

$$TAW = D \times WHC \tag{6.1}$$

where:

D is the effective root zone depth.

The lower the available moisture content of the soil, the deeper the roots of the crop must reach to secure a sufficiently large soil reservoir capacity. Under water harvesting conditions, the depth of soil wetting is usually limited, which limits the effective root zone depth (Table 6.2). Following a heavy rainstorm water may percolate below the root zone, where it will be out of the reach of plant roots. Plants can take up only

Table 6.1 Typical values for water holding (storage) capacity for different soil textures.

Soil	*Water holding capacity mm/cm*	*Soil*	*Water holding capacity mm/cm*
Coarse sand	0.2–0.6	Fine sandy loam	1.4–1.8
Sand	0.4–0.9	Loam, silty loam	1.7–2.3
Loamy sand	0.6–1.2	Clay loam, silty clay loam	1.4–2.1
Sandy loam	1.1–1.5	Silty clay, clayey silt	1.3–1.8

Source: Jensen (1980).

Table 6.2 Indicative values for maximum effective root zone depth for different groups of crops under water harvesting conditions.

Crop	*Rooting depth (m)*	*Crop*	*Rooting depth (m)*
Cereals (e.g. barley)	0.80	Bushes (e.g. *Atriplex spp.*)	0.75
Legumes (e.g. vetch)	0.60	Trees (e.g. olive)	1.20
Range grasses	0.40	Vegetables (e.g. tomato)	0.80

the available water retained within the root zone of crop. However, more often the soil depth is not sufficient to allow roots to grow to their full potential. Additionally, compacted, more or less impervious soil layers may act as physical barriers to root penetration (Rwehumbiza *et al.*, 2000).

6.2.1.3 *Infiltration rate*

Infiltration is the process whereby water enters the soil through the soil surface. Infiltration rate is a dynamic property varying with season and management. The main factors affecting infiltration rate are the soil type, the condition of the soil surface, and soil water content (Figure 6.1).

Structure and bulk density of the soil influence infiltration rate because of their relation to pore size distribution. Different soil types have different infiltration capacity rates. Figure 6.1 shows the infiltration characteristics of two types of soil. The loamy sand has a greater infiltration rate than silty clay. Silty clay may have a lower infiltration rate but higher runoff efficiency than loamy sand. Hence, care should be taken when designing a water harvesting system to assess what percentage of rainfall will be available as runoff water, how much of it will infiltrate, and how much will be used by the plants as soil moisture. A catchment (runoff area) with a final infiltration rate of 5 mm/h or less is a very good medium for harvesting rainwater (Eger, 1986). Table 6.3 presents typical values for final (basic) infiltration rates for various soils.

Soil surface cover affects runoff and infiltration. Large amounts of vegetation retard runoff and enhance infiltration.

6.2.2 Crop water requirements

In designing water harvesting systems, it is necessary to assess the water requirements of the crops intended to be grown (Allen *et al.*, 1998). However, it should

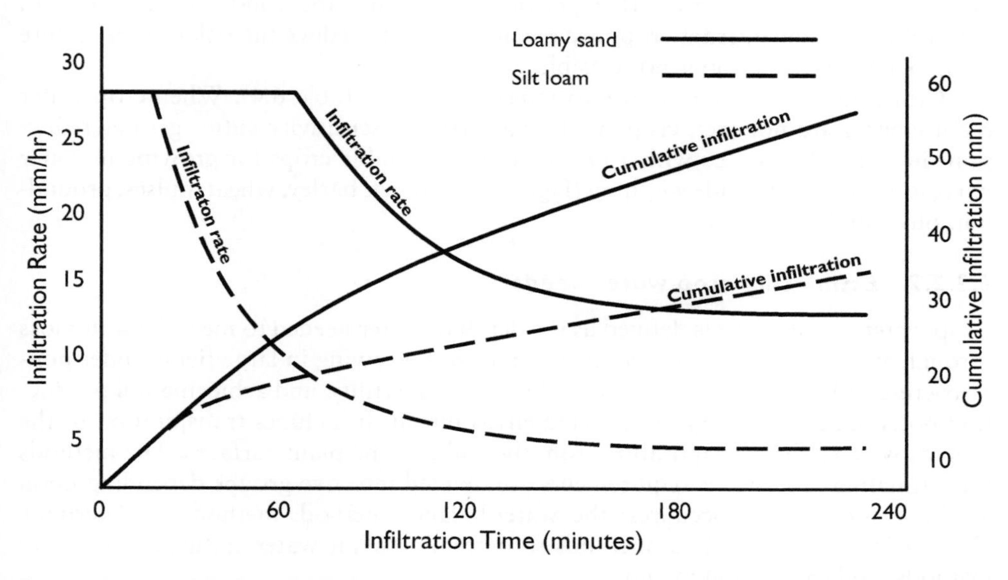

Figure 6.1 Typical infiltration rates and cumulative infiltration curves for loamy sand and silty clay soils under the same steady rain intensity of about 28 mm/h. (authors).

Table 6.3 Typical values for final (basic) infiltration rate for various soil textures.

Soil	*Final infiltration rate (mm/h)*	*Soil*	*Final infiltration rate (mm/h)*
Coarse sand	>22	Silty clay loam	9
Fine sand	>15	Clay loam	7.5
Fine sandy loam	12	Silty clay	5
Silt loam	10	Clayey soil	4

Source: Israelsen *et al.*, 1980.

be emphasized that water consumption by crops under water harvesting conditions (without interim storage) is different than that under fully irrigated agriculture. The latter is usually characterized by high production inputs and proper irrigation scheduling and management. Under water harvesting without storage outside the soil matrix, however, there is no irrigation scheduling per se.

6.2.2.1 Plant and drought

The objective of any water harvesting system in agriculture is to deliver a specific amount of water to the target (normally a crop). It is not always necessary to meet the full water needs of the crop; the objective should be to deliver an amount of water that results in an economical return. There is no control on the timing of water supply to the crop root zone under water harvesting conditions. Therefore, it is always necessary to select the crops that perform well under the conditions anticipated. Generally, drought-tolerant crops are recommended to reduce the risk of crop failure if supplemental irrigation is not feasible.

Crops differ in their response to moisture deficit (Table 6.4). When crop water requirements are not met, crops with high drought sensitivity suffer greater reductions in yields than crops with a low sensitivity. Suitable crops for growing in water harvesting systems include sorghum (Figure 6.2), millet, barley, wheat, pulses, groundnut, olive, and pistachio.

6.2.2.2 Estimating crop water needs

Crop water requirement is defined as the depth of water needed to meet the water loss through evapotranspiration of a disease-free crop, growing in large fields under non-restricting soil conditions, including soil water and fertility, and achieving full production potential under the given growing environment. It includes transpiration of the crop as well as direct evaporation from the soil and the plant surfaces. The methods for estimating crop water requirements are divided into five groups depending upon available tools and procedures: the water-balance method; methods to determine changes in soil water status; methods to determine plant water status; comparative methods; and calculation methods.

The water-balance method is frequently used for determining crop water requirements. A water balance can be established for various periods ranging from few days to one year. It can be implemented at various levels or scales, ranging from an entire basin to a plot of only few square meters, or even a lysimeter.

For a large area the total yearly evapotranspiration can be estimated as the difference between inflow (mainly precipitation) and outflow (mainly surface runoff) with the assumption of no change in the water storage of the soil profile.

Methods used to determine changes in soil water status include gravimetric methods, tensiometers, time domain reflectometers, etc.

Methods used to determine changes in plant water status include visual analysis of water stress indicators such as plant color and plant movements; measurement of infrared radiation; and direct measurement of relative water content, stomatal opening, transpiration rate, and osmotic and water potential of plants.

Table 6.4 General sensitivity of crops to drought.

Group	*Sensitivity to drought*	*Crop*
Group 1	Low sensitivity (well suited to water harvesting)	Barley
	↓	Safflower
		Millet
		Groundnut
Group 2		Sorghum
		Cotton
		Sunflower
Group 3		Beans
Group 4	High sensitivity (less suited to water harvesting)	Maize

Source: After Critchley & Siegert (1991).

Figure 6.2 Sorghum is well suited to cultivation under water harvesting conditions. Photo courtesy D. Prinz/Karlsruhe University, Germany.

Comparative methods include the use of evaporation pans and evaporimeters to determine the crop water requirements.

Calculation methods (Blaney-Criddle, Pan evaporation, Penman, Penman-Monteith, etc.) determine crop water requirements by using climate, soil, and crop-specific information. All available local data from experiments and existing water harvesting and irrigation schemes should be collected and evaluated prior to calculating crop and irrigation requirements. Daily values of crop water requirements, calculated according to the Penman-Monteith formula, are made available on the internet by a number of national meteorological services. Water needs vary from crop to crop. Trees require water all year round, while vegetables and cereals require water for only short growing seasons.

Details about these methods regarding procedures and calculations can be found in Allen *et al.*, (1998) and Katerji and Rana (2011). The FAO models CLIMWAT/CROPWAT are well suited to calculate crop water requirements for various crops for a range of climatological situations worldwide (www.fao.org/nr/water/infores_databases_climwat). These methods and models are, however, more applicable to intensive irrigated agriculture with high inputs and proper control and management of irrigation water. Less information is available on crop and tree water requirement under drought conditions prevailing in water harvesting projects. 'AquaCrop', a model developed by FAO for crop yield prediction under water deficit conditions, might be a useful tool also for agricultural water harvesting, with or without supplemental irrigation (Steduto *et al.*, 2008; www.fao.org/nr/water/infores_databases_aquacrop). Moreover, AquaCrop uses a relatively small number of parameters and input variables requiring simple methods for their determination. This is an advantage as availability of data is often limited in dry areas due to the low density of meteorological and agricultural research stations. The use of the 'consumptive crop water use' concept is more valid here than the 'crop water requirements', concept, because the former takes into account all factors and conditions, such as water stress, poor soil and fertility management, or inappropriate farming conditions, that influence the water use of a crop under given farming conditions. Moreover, availability of data is often limited in dry areas due to the low density of meteorological stations.

6.2.2.3 Field water budget

The concept of field water budget or balance is extremely important in evaluating the interseasonal behavior of the soil–water–climate–crop system. It is also helpful in evaluating the effectiveness of any water harvesting system.

In this approach the soil profile is subdivided into a number of layers. These layers may possess different physical properties and levels of soil moisture. The depth of water in each layer is determined by multiplying its volumetric water content by the layer thickness. The total water available in the root zone, the so called soil reservoir, is calculated by summing the depth of water in all the layers. The water available in the root zone is used as an indication to the status of the system or as a guide for timing and amount of irrigation application.

The basic form of the water budget equation is:

$$W_e = W_b + W_r + W_h - W_{et} - W_d - W_f \tag{6.2}$$

where:

- W_e is the total depth of water in the effective depth of root zone at the end of the selected time period
- W_b is the total depth of water in the effective depth of root zone at the beginning of the time period
- W_r is the total depth of rainfall during the period
- W_h is the total depth of water harvesting received during the period
- W_d is the total depth of deep percolation losses below the effective rooting depth of the crop
- W_{et} is the total depth of evapotranspiration during the period
- W_f is the total equivalent depth of surface runoff (spill) out of the cropping area (if any).

The term W_h in equation (6.2) depends, among many other factors, on the size (area) of the catchment relative to the cropped area (the target). There will be periods when the amount of precipitation is significantly less than the mean value used in the system design. While it is technically feasible to enlarge the catchment and storage in anticipation of below-average precipitation, it is not usually feasible to design a water-harvesting system to meet the least amount of precipitation that can be expected. The field water budget concept can be used to decide on the degree of risk that can be accepted regarding insufficient precipitation during some periods.

The final sizes (area) of the catchment and target should be determined using an incremental (weekly or monthly) water budget of water collected versus water requirements. This is to reduce the possibility that there will be periods of insufficient water.

In the absence of any measured climatic data, estimates of water requirements for common crops can be used. The average water requirements and growing periods of common crops and plants under irrigated conditions with proper management are given in Table 6.5. Main factors influencing crop water requirements include climate and crop type.

Table 6.5 Water requirements of selected crops under fully irrigated conditions with proper management.

Crop	*Growing period (northern hemisphere)*	*Water requirement (mm)*
Cash or oil crops		
Castor bean	April–Nov.	1130
Cotton	April–Nov.	1050
Flax	Nov.–June	795
Safflower	Jan.–July	1150
Soybean	June–Oct.	560
Sugar beet	Oct.–July	1090
Fodder crops		
Alfalfa	Feb.–Nov.	2030
Bermuda grass	April–Oct.	1100
Small-grain crops		
Barley	Nov.–May	635
Sorghum	July–Oct.	645
Wheat	Nov.–May	655
Fruits		
Grapefruit	Jan.–Dec.	1215
Grape (early maturity)	March–June	380
Orange (navel)	Jan.–Dec.	990
Vegetables		
Broccoli	Sept.–Feb.	500
Cabbage (early)	Sept.–Jan.	435
Cabbage (late)	Sept.–March	620
Cantaloupe (early)	April–July	520
Cantaloupe (late)	Aug.–Nov.	430
Carrot	Sept.–March	420
Cauliflower	Sept.–Jan.	470
Lettuce	Nov.–May	215
Onion (dry)	Sept.–Jan.	445
Onion (green)	Feb.–June	620
Corn (sweet)	March – Jun.	500
Green-manure crops		
Guar	July–Oct.	590
Peas (papago)	Jan.–May	495

Source: Erie *et al.* (1982).

For water harvesting system design, it is important to consider the timing of the water needs during the growing season. There are two stages or periods in which the availability of water is critical to the growth and production of field crops. The first stage is plant establishment, during which it is critical that the soil around the seed stays moist for a sufficient length of time to allow the seed to germinate and initiate root elongation. This activity takes place in the surface soil layer, where soil water is lost through evaporation. Keeping this layer wet usually requires some form of watering at two- to four-day intervals, depending on climatic and soil conditions. This watering period is very important with any annual crop that requires new plant establishment each growing season. Water requirement during establishment is very difficult to meet using a direct runoff water harvesting system that supplies water only during precipitation events. The second critical period is during the reproductive stage. At this stage, plants have usually developed a root system capable of extracting stored water from deeper layers in the soil profile. For winter barley in West Asia, for example, most of the water need occurs in March and April when the crop is in a stage of maximum growth and seed development. These water needs must be satisfied if the crop is to produce satisfactorily (Figure 6.3).

Figure 6.3 Stressed (right) and unstressed (left) barley crop during the reproductive growing stage under Mediterranean climate. The unstressed barley got some supplemental irrigation during critical periods. Photo courtesy ICARDA.

The crop water requirements presented in Table 6.5 are for ideal growing conditions with no biotic and abiotic stresses. These values must be adjusted under rainwater harvesting conditions. One simple and practical way to adjust these figures is by multiplying them by a stress factor (less than one) that depends on the economical yield level of the crop. For example: if we assume for a given crop that relative yield (i.e., actual yield/maximum yield) equals relative evapotranspiration, ET (i.e., actual ET/maximum ET) and the production costs and sale price allow to keep the production feasible with 50% of the maximum yield then a water supply of half of the value in Table 6.5 will do. There is a great need for research to determine appropriate measures to counteract such stress factors which are certainly site and crop specific. Tables 6.6 through 6.11 present crop consumptive use (actual water

Table 6.6 Consumptive water use of vegetables in Jordan under water harvesting conditions.

Month	*Pan-A evaporation (mm)*	*Potential* ET_0† *(mm)*	*Consumptive water use (mm)*
March	135	94	28
April	227	158	63
May	275	193	115
June	302	212	106

Note: †Reference evapotranspiration. Alfafa is the reference crop.
Source: Drolett *et al.* (1997).

Table 6.7 Consumptive water use of legumes in Jordan under water harvesting conditions.

Month	*Pan-A evaporation (mm)*	*Potential* ET_0† *(mm)*	*Consumptive water use (mm)*
Dec.	114	80	31
Jan.	82	58	35
Feb.	103	73	67
March	135	94	104
April	227	158	53

Note: †Reference evapotranspiration. Alfafa is the reference crop.
Source: Drolett *et al.* (1997).

Table 6.8 Consumptive water use of cereals in Jordan under water harvesting conditions.

Month	*Pan-A evaporation (mm)*	*Potential* ET_0† *(mm)*	*Consumptive water use (mm)*
Dec.	114	80	36
Jan.	82	58	47
Feb.	103	73	87
March	135	94	113
April	227	158	53

Note: †Reference evapotranspiration. Alfafa is the reference crop.
Source: Drolett *et al.* (1997).

Table 6.9 Consumptive water use of range grasses in Jordan under water harvesting conditions.

Month	*Pan-A evaporation (mm)*	*Potential* ET_0[†] *(mm)*	*Consumptive water use (mm)*
Feb.	103	73	5
March	135	94	38
April	227	158	95
May	275	193	97
June	302	302	14

Note: [†]Reference evapotranspiration. Alfafa is the reference crop.
Source: Drolett *et al.* (1997).

Table 6.10 Consumptive water use of range bushes in Jordan under water harvesting conditions.

Month	*Pan-A evaporation (mm)*	*Potential* ET_0[†] *(mm)*	*Consumptive water use (mm)*
Jan.	82	58	6
Feb.	103	73	8
March	135	94	19
April	227	158	47
May	275	193	58
June	302	212	63
July	310	218	44
Aug.	274	192	38
Sept.	237	166	17
Oct.	119	139	15
Nov.	169	119	12
Dec.	114	80	9
Total	2347	1702	342

Note: [†]Reference evapotranspiration. Alfafa is the reference crop.
Source: Drolett *et al.* (1997).

Table 6.11 Consumptive water use of fruit trees in Jordan under water harvesting conditions.

Month	*Pan-A evaporation (mm)*	*Potential* ET_0[†] *(mm)*	*Consumptive water use (mm)*
Jan.	82	58	12
Feb.	103	73	15
March	135	94	20
April	227	158	32
May	275	193	38
June	302	212	42
July	310	218	44
Aug.	274	192	38
Sept.	237	166	34
Oct.	119	139	28
Nov.	169	119	23
Dec.	114	80	16
Total	2347	1702	342

Note: [†]Reference evapotranspiration. Alfafa is the reference crop.
Source: Drolett *et al.* (1997).

use under prevailing growing conditions) for various crop groups grown in Jordan under typical water harvesting conditions, which are notably less than the values presented in Table 6.5. For example, the consumptive use for cereals in Table 6.8 is nearly half of the values in Table 6.5.

To secure the livelihood of pastoralists in dry areas, agroforestry systems, i.e., the protection/cultivation of fodder trees and bushes in close association with pasture crops, are essential. Water harvesting systems can contribute to secure an ample water supply for both elements (Droppelmann & Berliner, 2003).

6.3 RAINFALL

The quantity of rainfall that might occur during a given time period is one of the most difficult parameters to predict. This is particularly true in dry areas, which are characterized by low, non-uniform, and erratic rainfall that mostly comes in unpredictable, sporadic storms.

6.3.1 Inter-seasonal distribution of rainfall

Monthly precipitation data are usually available for most places. Averages derived from long-term precipitation records are the most common form available. However, in most arid and semi-arid regions there are wide, random fluctuations from the mean for any given month. These fluctuations can have a major impact on the ability of a water harvesting system to supply the required water. To minimize the effect of variations in precipitation, it is recommended to use records covering more than 10 years. If there are extreme variations in precipitation, data from the two wettest years should be eliminated when computing mean values.

With adequate long-term data, probability analysis can be used to estimate likely monthly precipitation for use in designing the optimum system. For example, Table 6.12 and Figure 6.4 present the results of monthly rainfall analysis for Breda, northern Syria. On average, cumulative precipitation exceeds barley water use until mid–March, after which crop water use exceeds cumulative precipitation. Heading, flowering, and grain filling occur from mid-March onwards, a period when if little or no rain falls. Without supplemental irrigation, crop production will be substantially reduced due to severe moisture deficit and stress.

6.3.2 Design rainfall

The design rainfall for water harvesting is determined by frequency analysis of the available rainfall data. The analysis may be performed on yearly, monthly, or weekly rainfall series (Figure 6.4). First, an acceptable exceedance probability, i.e., the percentage of years, months, or weeks in which rainfall will exceed a given value, is selected; this is commonly between 60 and 80% in designing water harvesting systems. Using the long-term rainfall data, this exceedence value can be translated into a rainfall depth, and this is taken as the design rainfall. These probabilities give dependable rainfall amounts that are less than mean rainfall values.

Table 6.12 Monthly average, wettest, and driest year rainfall at Breda, northern Syria (1980–1999) along with the monthly rainfall for the 70% probability of exceedance year and mean monthly barley water use in the area. No rainfall during June-Sept. period.

Month	*Average (mm)*	*Wettest year (1987/88) (mm)*	*Driest year (1989/90) (mm)*	*70% year (1997/98) (mm)*	*Barley water use*
Oct.	18	68	25	35	0
Nov.	33	36	28	11	20
Dec.	38	41	34	29	25
Jan.	49	86	37	47	35
Feb.	44	82	43	17	50
March	37	66	8	39	85
April	23	19	3	39	95
May	16	10	6	9	0
Total	258	408	184	226	310

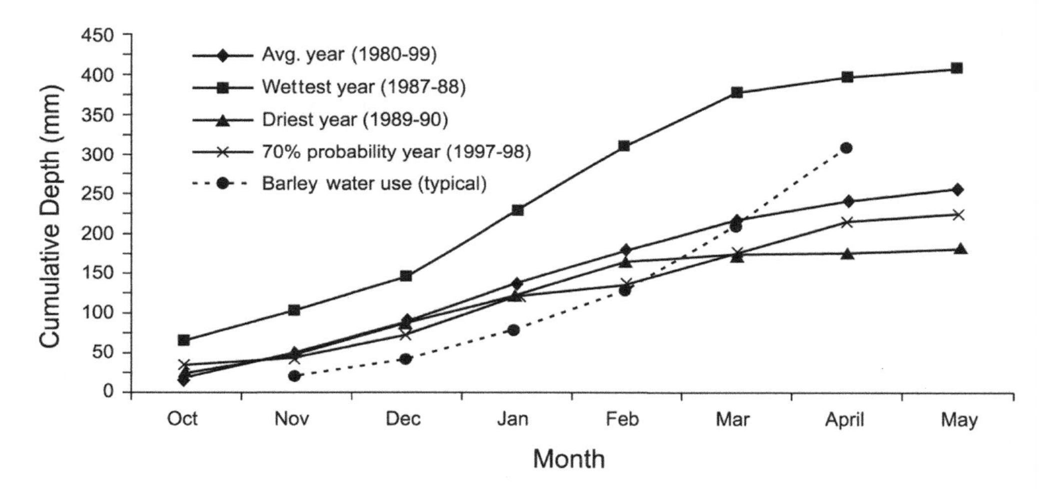

Figure 6.4 Cumulative rainfall and cumulative water use by barley at Breda, Syria, 1980–99.

A probability analysis can be carried out for a particular month during the growing season. Of course, monthly rainfall varies and fluctuates more than yearly total rainfall. Also, daily rainfall exhibits more variation than weekly or monthly rainfall. Therefore, it is not acceptable to carry out probability analysis for each month separately then take a certain exceedance probability rainfall for all the months and consider their total as the yearly rainfall for that exceedance probability. It can be shown, that for any exceedance probability greater than 50%, the sum of the probable monthly rainfall of all the months is greater than the probable yearly rainfall for the same exceedance probability. This is because:

$$\sum_{i=1}^{N}(\bar{X}_i + KS_i) = \sum_{i=1}^{N}(\bar{X}_i) + K\sum_{i=1}^{N} S_i = \bar{X} + K\sum_{i=1}^{N} S_i > \bar{X} + K\, S_y \qquad (6.3)$$

where:

$\bar{X}_i$ is the average monthly rainfall
N is the number of rainy months in the year
K is the frequency distribution factor, which depends on the probability level
S_i is the standard deviation of the monthly rainfall series. The size (i.e., number of values) of each monthly series equals number of years in the records
$\bar{X}$ is the mean annual rainfall
S_y is the standard deviation of the annual rainfall series.

Equation (6.3) verbally states, that the sum of the standard deviations of monthly rainfall series is greater than the standard deviation of the sums of the monthly rainfall (i.e., yearly rainfall). Figure 6.4 shows the monthly rainfall based on frequency analysis for yearly rainfall and not for monthly rainfall series analysis. The Figure indicates that there is no water shortage to barley, at this site, before the beginning of March. Figure 6.4 shows that based on 70% rainfall probability, there is shortage of about 50 mm of water to barley by the end of April. Barley at Breda usually does not need water after the end of April. Thus, the design depth of harvested water that should be furnished by the harvesting system should be around 50 mm. Later, this design parameter will be discussed further in this chapter.

6.3.3 Need for storage

In rainfed agriculture the intraseasonal distribution of rainfall is as important as the total rainfall. If the bulk of the rain always or mostly falls long before the critical growth stages of the crop (when crop water demand is highest) a storage facility other than the soil profile will be needed. The stored water will be used for supplemental irrigation during prolonged rainless periods.

In many installations, there will be several combinations of catchment and storage sizes that will provide the required quantities of water. The cheapest system will often be the one with the smallest storage volume. In these instances, it is important to determine the minimum storage volume that will provide sufficient water to meet crop needs during periods of low rainfall. At other times, the storage may overflow (Oweis & Taimeh, 2001). While it may appear that water is being wasted, there is no need for greater storage capacity if there are no periods when there is insufficient water. Less storage capacity is needed if the periods of maximum precipitation coincide with the periods of maximum use. More storage capacity is needed if periods of greatest precipitation occur either long before or after the periods of greatest water need.

6.3.4 Basic design procedure

There is no standard or ideal design for a water harvesting system. Each site and water use has unique design requirements and each system must be fitted to the local conditions and needs. There are many separate elements that must be considered in designing of water harvesting systems, including precipitation patterns, water requirement patterns, alternate water sources, soil types, land topography, available materials, labor, and acceptability of water harvesting concepts to the water users. Many of these factors are interrelated and it is difficult to assess their role separately; hence they must be simultaneously considered during the design and implementation of a water harvesting system.

The design procedure depends mainly on the type of crop and water harvesting technique. Long-slope and floodwater systems usually involve the design of a small dam, diversion structures, and water conveyance and/or distribution systems. They may also include facilities to store water for subsequent use. The size of a long-slope or floodwater catchment area is often not under the designer's control. Thus, the designer determines the extent of the cropped area that may be served with the expected runoff.

Runoff in macrocatchment systems may be determined directly using gauges (Figure 6.5), calculated from flow cross sectional area and velocity measurements, or estimated using simulation models. Such large-scale systems are beyond the ability of individual farmers, and advice should be sought from engineers.

Unlike the macro-catchment design, the most important element in the design of micro-catchments is the determination of the catchment area required to supply the required amount of water to the target area. The size of the catchment area will depend on rainfall characteristics, land slope, soil characteristics, land cover, crops, and economics. In macro-catchment systems, water harvesting structures are important, that will convey, store, and/or distribute the collected water to the crops or other uses. On-farm micro-catchment water harvesting systems do not require water-storage structures, because the harvested water is stored in the soil of the crop's root zone.

The design of the system must be flexible so that any necessary change or modification in the cropped area or crop type during implementation or in the future may be accommodated without undue difficulties.

Figure 6.5 Runoff can be measured directly using gauges, like this one in Kanguessanou, Kayes Province, Mali. Photo courtesy W. Klemm/Karlsruhe University.

Emphasis in the following chapter is on the design of micro-catchment water harvesting systems.

6.3.5 Selection of site and method

Not all areas are suitable for implementing water harvesting systems. Apart from basic technical requirements, the selected technique must be compatible with local social and farming systems. Table 6.13 outlines the requirements of the most important mico-catchment water harvesting techniques. This information can be helpful in selecting the most suitable one.

The intended use of the water harvested (domestic, consumption by livestock, irrigation of crops, or multipurpose) determines to some extent the site and the most appropriate methods. Topography is a major factor in selecting an appropriate water-harvesting technique. Water harvesting systems, however, may be implemented on a wide range of slopes. Generally, steeper slopes are used as catchments and shallower slopes or flat land is used for cropping. Soils are generally shallower on steep slopes and deeper on shallower slopes, which is advantageous to the two uses.

Soils with high infiltration rates, such as sandy soils, are not good as catchment areas for water harvesting unless runoff inducement measures are applied. Higher investments in runoff inducement are only economic, when the harvested water is used for human and animal consumption or for the production of high-value crops. Both soil texture and depth influence the total water-storage (holding) capacity of the soil profile, which, in turn, controls the amount of water that can be made available to crops during dry periods.

Table 6.13 Outline requirements for major micro-catchment water harvesting techniques.

Technique	*Preferred slope*	*Catchment:cropping area ratio (CCR)*	*Type of crop*	*Description*
Inter-row water harvesting	0–5%	1:1–5:1	Field crops, vegetables, trees	Runoff area: smoothing, compaction; run-on-strips: about 2 m wide
Semi-circular bunds	1–5%	4:1–8:1	Trees, range, fodder, field crops	4–8 m diameter, 5 m vertical distance
Vallerani water harvesting	2–10%	6:1–10:1	Bushes and trees	Distance between plow lines: 4 m
Negarim	1–5%	1:1–25:1	Trees and bush crops	Squares of 3 × 3 m to 10 × 10 m
Meskat	2–15%	1:1–5:1	Trees and bush crops, field crops	Spillways needed
Contour bunds	1–25%	5:1–10:1	Trees and bush crops, range, field crops, vegetables	5–10 m between rows
Small pits ('Zay')	0.5–5%	1:3	Field crops, range	Diameter of pits: 20–30 cm
Contour bench terraces	15–65%	1:1–10:1	Trees and bush crops	Small terraces 3–10 m apart

Water rights, land tenure, and land use may limit the choice of site and technique. Overlooking these issues has contributed to the failure of many water harvesting projects. Collective land ownership increases the number of options, including macro-catchments. Large-scale systems may be more economical and require less individual installation and maintenance.

Selection of best-suited micro-catchment method

Soil depth and slope angle are key factors influencing choice of water harvesting technique. For example, contour bench terraces are recommended on steep slopes exceeding 40%, while contour bunds are not recommended for slopes greater than 25% (Figure 6.6).

Social attitude towards private ownership, individualism, and personal interests are changing in many countries, and small individual farms may be better suited to

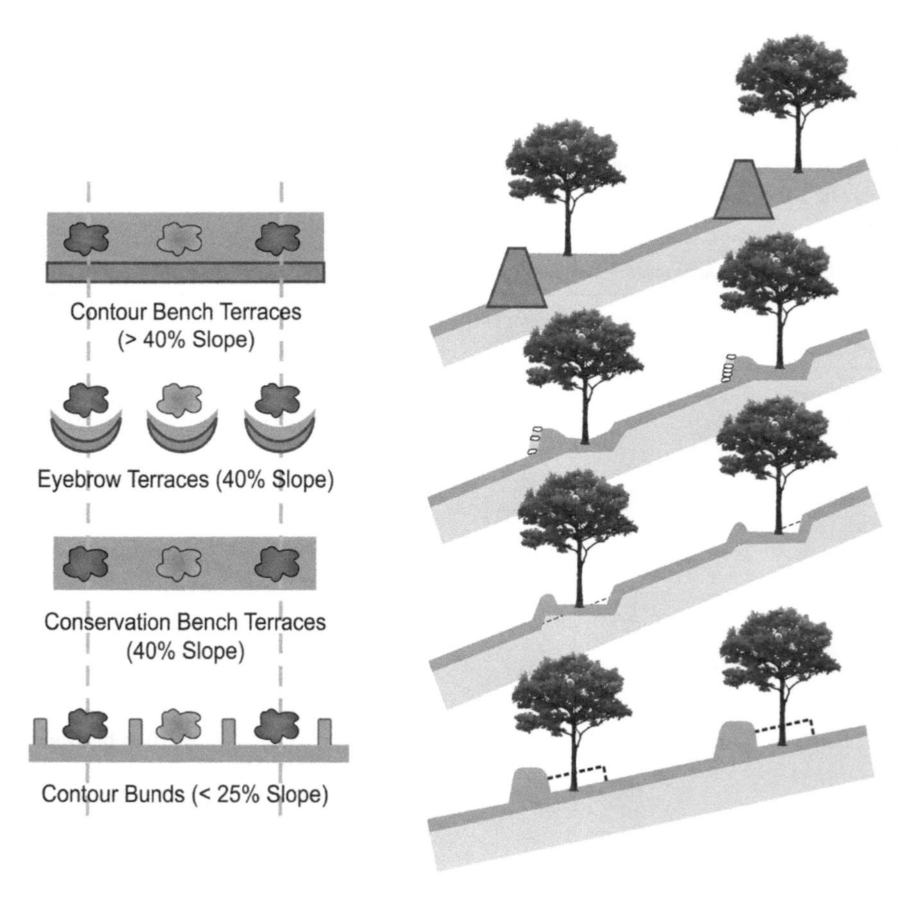

Figure 6.6 Angle of the slope is a key factor influencing the choice of water harvesting technique. (Prinz, 2010).

micro-catchment systems than are collectively owned land. The farmer's ability to operate and maintain the system determines the level of sophistication of the system that can be employed. Construction requirements such as availability of materials and skilled labor also influence the choice of system.

6.3.6 Selection of crops

In addition to the climatic and local farming conditions, the following should be considered in selecting crops:

- Drought-tolerant crops reduce the risk of crop failure.
- Annual cops and trees already grown in the local farming system are ideal.
- In drier environments, shrubs and trees are more likely to regrow after browsing and harvesting than are grasses and legumes.
- Seasonal crops that grow during the rainy season should always be given priority in order to assure quick payoff from the water harvesting system.
- Soil depth influences water availability. For example, young trees on shallow soil may not survive through a long, hot, dry summer preceded by an erratic rainy

Table 6.14 Preferred climatic zone, drought and water-logging tolerance of trees.

	Rainfall		
Tree species	*Semi-arid/semi-humid 500–900 mm/year*	*Arid/semi-arid 150–500 mm/year*	*Tolerance to temporary water logging*
Acacia albida (syn. Faidherbia albida)	Yes	Yes	Yes
A. nilotica	Yes	Yes	Yes
A. saligna	No	Yes	Yes
A. senegal	Yes	Yes	No
A. seyal	Yes	Yes	Yes
A. tortilis	Yes	Yes	No
Albizia lebbeck	Yes	No	No
Azadirachta indica (syn. Antelaea azadirachta)	Yes	No	Some
Balanites aegyptiaca	Yes	Yes	Yes
Cassia siamea (syn. Senna siamea)	Yes	No	No
Casuarina equisetifolia	Yes	No	Some
Colophospermum mopane	Yes	Yes	Yes
Cordeauxia edulis	No	Yes	?
Cordia sinensis	No	Yes	?
Delonix elata	Yes	No	?
Eucalyptus camaldulensis	Yes	Yes	Yes
Prosopis chilensis	Yes	Yes	Some
Prosopis cineraria	Yes	Yes	Yes
Prosopis juliflora	Yes	Yes	Yes
Ziziphus mauritiana	Yes	Yes	Yes

Source: Critchley and Siegert (1991).

season (such as in the Mediterranean climate). If the rainy season is much below average, even established trees may not survive on shallow soil. Choice of crop should be based on indicative values for maximum effective root zone depth under water harvesting conditions (see Table 6.2).
– Longer periods of waterlogging may occur; hence, crops sensitive to waterlogging should be avoided.

6.3.7 Runoff estimation

The likely amount of runoff that can be harvested can be estimated from the following equation:

$$r = R \times RC \qquad (6.4)$$

where:

r is the runoff depth
R is the rainfall depth
RC is the runoff coefficient, usually less than 0.5 (natural conditions with or without minor treatment) (see equation (2.5) for calculation of RC).

6.3.8 Catchment: Cropping area ratio (CCR)

The ratio of the catchment size to the cultivated area (the target) (CCR) represents the degree of concentration of rainfall in water harvesting systems. Accurate basic data are needed to determine this ratio accurately. Overestimating the ratio needed wastes land and water.

Not all the water harvested is beneficially used by the crops in the cultivated area. Some of it evaporates, and some of it percolates below the effective root zone of the crop, putting it beyond the reach of the crop roots. There are two main reasons for deep percolation: (a) too little water-storage capacity in the root zone especially during periods of heavy and/or frequent rainfall; (b) nonuniformity of water distribution across the cropped area. All these factors must be allowed for in determining the CCR through a storage efficiency factor. This efficiency is defined as the ratio of the volume of harvested water that is stored in the effective depth of the crop root zone to the total volume of water harvested. Typical values for storage efficiency range from 60% to 80%.

Seasonal volume of water harvested must equal the gross seasonal volume of extra water required to satisfy crop consumptive use.

The volume of water harvested can be calculated from the following equation:

$$\text{Volume of water harvested} = A \times R \times RC \qquad (6.5)$$

where:

A is the catchment area
R is the design rainfall (seasonal or yearly)
RC is the design seasonal or yearly runoff coefficient

The gross volume (GVE) of extra water required to satisfy crop consumptive use can be calculated from the following equation:

$$GVE = a\ (U - R)/E \qquad (6.6)$$

where:

a is the area cultivated or cropped
U is the seasonal or yearly crop consumptive use
E is the storage efficiency of water in the effective root zone depth of the cropped area.

By rearranging equations (6.5) and (6.6), the ratio (A/a) that represents the ratio of the catchment area to the cropped area (CCR) can be solved as follows:

$$A/a = (U - R)/(R \times RC \times E) \qquad (6.7)$$

6.3.9 Design examples

Two simple design examples for illustrating the use of the procedure presented in the previous section are given below:

Example (1):

Crop: Barley

Crop consumptive use = 250 mm

Design rainfall = 139 mm (at 70% exceedance probability)

Runoff coefficient = 0.25

Storage efficiency factor = 0.80

Therefore, by using equation (6.7), the CCR (A/a) will be:

$$A/a = (250 - 139)/(139 \times 0.25 \times 0.8)$$

$$= 4$$

For the runoff-strip water harvesting technique, this ratio means that if the width of the cropped strip is 1.5 m, the width of the compacted catchment strip should be 6 m.

Example (2):

Same data given in example (1) except that the design rainfall is 114 mm at 90% exceedance probability.

Therefore, by using equation (6.7), CCR will be:

$$A/a = (250 - 114)/(114 \times 0.25 \times 0.80)$$

$$= 6$$

6.3.10 Optimization of system design

The two examples presented above show that increasing the dependability level of the water harvesting system from 70% (seven successful years and three years of failure out of 10 years period) to 90% (which reduces design rainfall) increased the area ratio from 4:1 to 6:1. Increasing the area ratio from 4:1 to 6:1 means reducing the cropped area from 0.2 ha/ha to 0.14 ha/ha of gross area of land. Therefore, in a successful year, the crop production per unit gross area (i.e., catchment plus cropped area) of land will decrease as the area ratio is increased. However, the number of success years per decade increases as the system dependability is increased. Two simplified assumptions are needed to pursue the present rational analysis further. First, the crop production per unit area is the same (i.e., constant) in any of the success years. Secondly, the limited crop (biomass) yield in a failure year just covers the production cost; therefore, the net economic return is zero.

Table 6.15 shows the effect of changing the exceedance probability on average crop production per unit area based on the above assumptions. The table indicates that 70% exceedance probability gives the highest average crop production per unit area.

6.3.11 Further considerations in area ratio selection

Determining the CCR on the basis of seasonal rainfall and seasonal crop water consumptive use has many problems and drawbacks, including the following:

- It does not take into consideration the dynamics of soil moisture in the root zone along the growing season.
- It overlooks the effect of soil water storage capacity on this dynamics. During relatively long and continuous rainfall periods, part of the infiltrated rainwater may percolate out of the root zone depth. The parameter E is introduced in equation (6.7) to take care of this issue, but the non-uniformity of the concentrated runoff water on the cropped area complicates the issue.
- More importantly, it does not give any consideration to the temporal distribution of rainfall over the growing season.

Table 6.15 Determination of the optimal area ratio for crop consumptive use of 250 mm, runoff coefficient of 0.40, and storage efficiency in the root zone of 0.50.

Exceedance probability (%) [1]	*R (mm)* [2]	*A/a* [3]	*a (ha/ha)* [4]	*Average crop production (units/ha)* [5]
50	156	3:1	0.25	0.125
60	147	3.5:1	0.22	0.132
70	139	4:1	0.20	0.140
80	125	5:1	0.17	0.136
90	109	6.5:1	0.13	0.117

Note: [3]: From equation (6.7).
[4]: $a = 1/(1 + A/a)$.
[5]: From the product of column [1] by column [4].

Determining a suitable value of CCR may be greatly improved by basing the assessment on shorter time periods, such as weeks or 10-day periods. Shorter time periods-based analysis allows more accurate monitoring of the dynamics of the system in-terms of soil moisture in the root zone of the crop and variations in rainfall. To illustrate this approach and to compare its results with that based on seasonal data, the first three columns in Table 6.16 present rainfall and estimated consumptive use for 10-day periods for a rainfall season in a dry area.

The calculations used in Table 6.16 can then be repeated for various values of the CCR. Figure 6.6 shows the result of such calculations. The 50% AW line is marked to show the readiness of the soil moisture in the root zone to crop. Soil water above this line is readily or easily available to the crop while below the line is not easily available. The figure shows that with a CCR of 1:1 no available water remains in the root zone after Day 96. With a CCR of 3:1 more than 40 mm of available water remains in the root zone at the end of the growing season. With a CCR of 1:1, half of the land is cropped, while with a CCR of 3:1 only one quarter of the land is cropped. However, the yield per unit area of cropped land is higher with a CCR of 3:1 than with a CCR of 1:1. Therefore, there is a trade-off between the size of the cropped area and the yield per unit of cropped area that demands careful study and further consideration.

Table 6.16 Depth of harvested water and available water in the soil of the cropped area for 10-day intervals with catchment: cropping area ratio (*CCR*) of 1:1, runoff coefficient (*RC*) of 0.50 and water distribution efficiency on the cropped area (*E*) of 85%.

Day	*10-day sum rain (mm) (R)*	*10-day consumptive use (mm)*[†]	*Harvested water (CCR × R × RC × E) (mm)*	*Sum of rain and harvested water (mm)*	*Available water (AW)(mm)*
0[*]	0	0			40[‡]
10	15	20	6.4	21.4	41[§]
20	5	20	2.1	7.1	29
30	0	20	0.0	0.0	9
40	25	10	10.6	35.6	34
50	15	10	6.4	21.4	46
60	0	10	0.0	0.0	36
70	10	25	4.3	14.3	25
80	20	25	8.5	28.5	28
90	5	25	2.1	7.1	10
100	10	30	4.3	14.3	–5
110	20	30	8.5	28.5	–7
120	15	30	6.4	21.4	–16
130	10	40	4.3	14.3	–41
140	10	35	4.3	14.3	–62
150	0	30	0.0	0.0	–92

Note: *Beginning of the growing season.
†Mean daily consumptive use for the 1st, 2nd, 3rd, 4th, and 5th 30-day interval is 2, 1, 2.5, 3, and 4 mm/day, respectively.
‡Available water in the root zone of the crop at the beginning of the growing season (from early-season rainfall).
§Calculated as: 40 mm available water + 15 mm direct rain + 6.4 mm harvested water – 20 mm consumptive use.

Figure 6.7 indicates that a CCR of either 2:1 or 2.5:1 is a better choice than 1:1 and 3:1. If the analysis shown in Figure 6.7 is performed for all years with available data, the CCR value that would give the best results in most years can be identified and selected for the design of the water harvesting system.

6.4 DESIGN CONSIDERATIONS FOR TREES

Water harvesting may not be successful for trees and other perennial crops unless the soil of the cropped area is deep enough and has sufficient water-holding capacity to cope with the evaporative demand of dry, hot, and rainless periods that may last longer than seven months (e.g. in the Sahel region). In the Mediterranean climate of the Near East the dry, hot, rainless period extends from June to September, inclusive. The crop must survive through this period solely on the moisture left over and stored in the root zone reservoir at the end of the rainy season. Ideally, the root zone reservoir should be full with total available water at the beginning of the rainless period. This point will be discussed further in this chapter and in Chapter seven.

6.4.1 Design for trees

Two main difficulties arise when designing water harvesting systems for trees.

First, it is difficult to determine the area exploited by the rooting system of a tree. This refers to the target area, a, in equation (6.6). One way to resolve this difficulty is

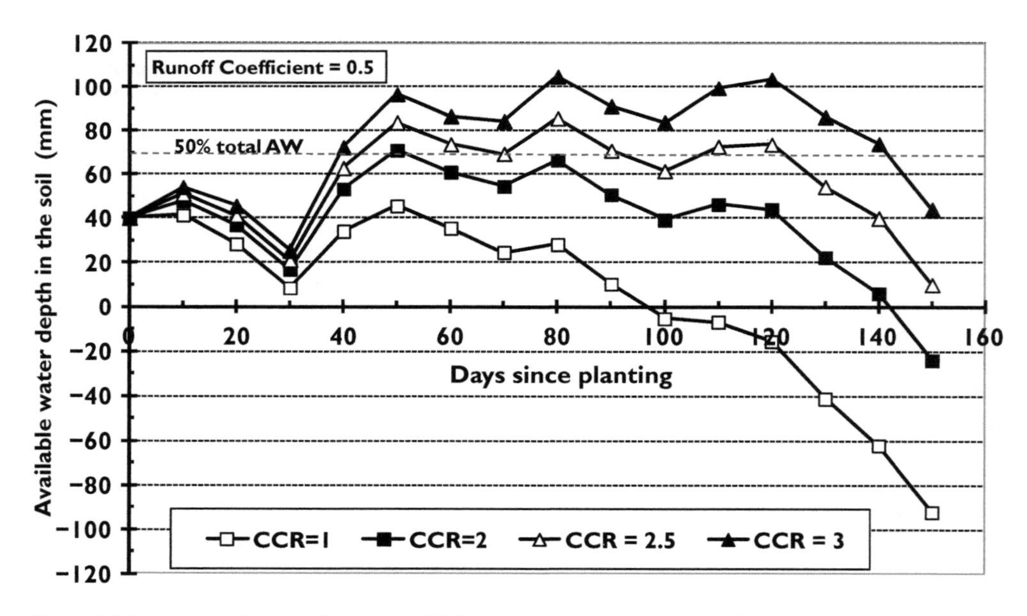

Figure 6.7 Variation of available water (AW) in the crop root zone under micro-catchment water harvesting over the growing season using different catchment: cropping area (CCR) values. Total seasonal rainfall 200 mm (40 mm of it before planting) and assumed crop water use of 360 mm.

to set a, equal to the area shaded by the canopy of the tree at noon time. This assumption is used in the design of drip-irrigation systems for trees. Further, the catchment area, A, is assumed to include the target area, a. The purpose of this assumption is to simplify the design and reduce geometrical complications, especially if the target area is not located at the corner or to one side of small runoff basins. In dry areas, A ranges between 10 and 100 m^2.

The second difficulty encountered in designing water harvesting systems for trees is the lack of information on the minimum (life-saving) and/or economical return level of water to be made available to the trees. Only rough estimates can be made for indigenous trees, such as pistachio, olive, apricot, and almond (Table 12 in Critchley & Siegert, 1991).

Equation (6.7) can be used to estimate the size of catchment area needed to supply the water need of trees. For example, for a *negarim* micro-catchment water-harvesting system, if the annual tree consumptive use (U) = 440 mm, design annual rainfall (R) = 200 mm, RC = 0.4, and E = 0.75, A/a = 4. If a (the area shaded by a tree) = 12 m^2, the total area of each *negarim* unit must be 48 m^2.

6.4.2 Life-saving harvested water

As was discussed earlier, trees and other perennials survive hot, dry periods that may last several months. Ideally the root zone reservoir should be full at the beginning of these periods. This can be achieved in a number of ways.

A small storage facility may be incorporated in the water harvesting system and the water stored applied to the cropped area to fill the root zone reservoir at the beginning of the dry periods. Storing the water in the root zone minimizes evaporation and seepage losses that usually occur in water storage reservoirs. Alternatively, the CCR may be increased to increase the volume of water harvested towards the end of the rainy season. However, this may put additional stress on the earthen bunds and dikes of the system. Therefore, these structures must be strengthened. It may be best to employ a strategy that combines both of these approaches.

6.5 DIMENSIONING, MATERIALS AND ESTIMATION OF QUANTITIES

The availability of suitable construction materials at or near the construction site is one of the basic requirements for a successful water harvesting project and influences the choice of water harvesting technique. Stones of a suitable size and structurally stable soils are the two basic construction materials for water harvesting systems. Manufactured construction materials can be used, but they come at a relatively high cost. Water harvesting is in general a low-cost technology that should depend on local natural resources including construction materials.

6.5.1 Dimensioning and system layout

After determining the CCR for the micro-catchment water harvesting system, the next step in implementation is to dimension and lay out the system on the ground.

Dimensioning refers to selecting the proper dimensions (spacing, length, width, height, etc.) of the various components of the water harvesting system needed to satisfy the design requirements of the system.

This process can be illustrated using the case of a crescent micro-catchment water harvesting system. Figure 6.8 shows a typical layout for such a system. The catchment may be defined as the space or area enclosed inside the shape bcdefgha. The area of this catchment A_c may be, for practical purposes, approximated by:

$$A_c = (Y - 0.5\,Z)\,S + Y \times W \qquad (6.8)$$

where:

Y is the mean distance between contours
S is the distance between adjacent crescent bunds
W is the width of the crescents (see Figure 6.8).

$$Y = Z + H \qquad (6.9)$$

where: Z and H are as defined in Figure 6.8.

A_t is the target (or cropped) area which may be defined by the segment of the circle in which the chord is the straight line connecting the two ends of the crescent. The area of this circle segment may be roughly approximated by:

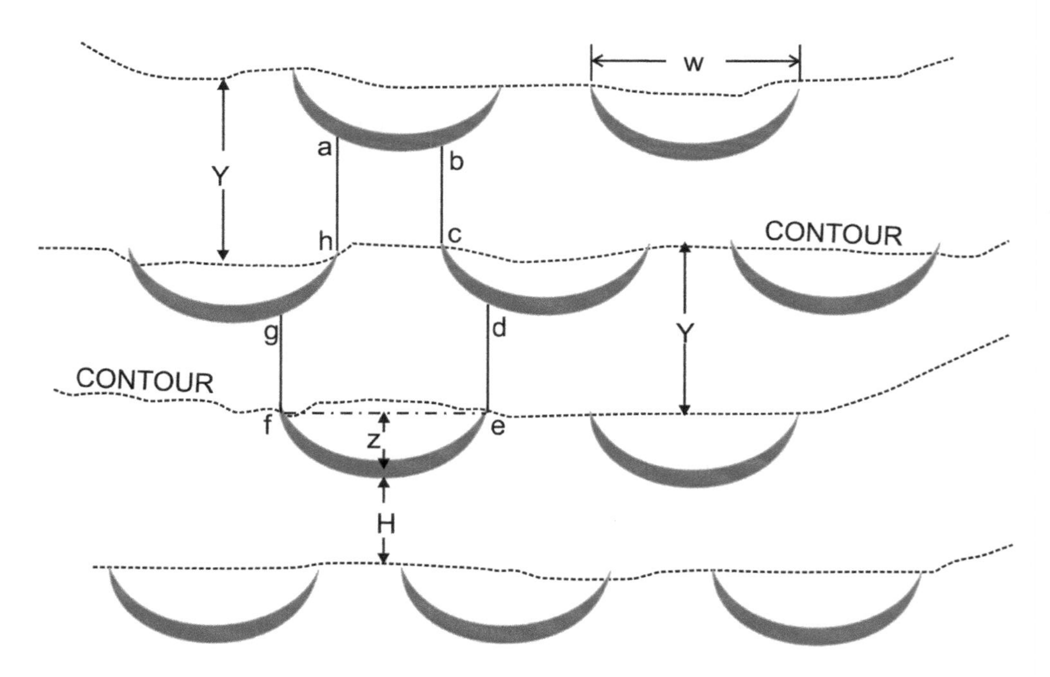

Figure 6.8 Typical layout of a crescent-type micro-catchment water harvesting system.

$$A_t = \pi W \times Z/4 \qquad (6.10)$$

The CCR, A_r, will be:

$$A_r = (A_c/A_t) = [(Y - 0.5\,Z)\,S + Y \times W]/(\pi W \times Z/4) \qquad (6.11)$$

For a semicircle, Z will be equal to W/2 and A_t in equation (6.7) will be equal to $\pi W^2/8$, which is the area of a half circle or a semicircle. Therefore:

$$A_r = [(Y - 0.25\,W) \times S + Y \times W]/(\pi W^2/8) \qquad (6.12)$$

Furthermore, if k = S /W, equation (6.12) becomes:

$$A_r = (8/\pi)\,[(Y/W)\,(k + 1) - (k/4)] \qquad (6.13)$$

To capture all runoff in the system, $k \leq 1$.

In Figure 6.7, if S = W, i.e., k = 1, then $Ar = (8/\pi)\,[(2Y/W) - 0.25]$

If k = 0, $A_r = (8/\pi)\,(Y/W)$

As an example for semicircular crescents, if k = 0, W = 5 m and H = 4 m (that is Y = 6.5 m), the CCR, Ar, is equal to 3.3. If this value is equal to the CCR (i.e., A/a) calculated by equation (6.7), the dimensioning and layout of the semicircular bunds are acceptable, otherwise a better layout should be looked for.

For contour ridges, dikes, and bunds, the height of the bund is more or less constant. However, for crescent, trapezoidal, rectangular, and triangular micro-catchment systems, bund height is not constant because of the slope of the ground. The height is greatest at the lowest point and least at the upper ends (tips) of these bunds. Therefore, care should be taken that the depth of impounded water behind (i.e., upper side of the) bund does not overrun it. This requires that the level of the bund top at the lowest point of the bund is at least 10–15 cm higher than the ground level at the upper ends or tips of the bund. This condition assures that excess harvested water runs off around the tips of the bunds rather than running over the bund. Newly constructed earthen bunds are apt to settle during the first season. Allowance for this settlement should be provided when deciding on the height of the bund. Moreover, the surface storage capacity of the bund should be sufficient to store the anticipated volume of runoff water calculated in equation (6.5).

Land slope plays a major role in the layout of the system. Land slope determines the distance between contours (Y in Figure 6.8). This in turn affects the boundaries and size of the catchment area. Sites with nearly uniform slopes are ideally suited to micro-catchment water harvesting because the contour lines are almost parallel. Nevertheless, careful judgment and some engineering sense are required to lay out water harvesting systems successfully. If the land slope on the site is not uniform, the planner should choose a representative slope that dominates in the site for dimensioning and laying out the system.

Figures 6.9–6.11 give construction details of a *negarim* on a 3% slope. The total size of the *negarim* is 5 × 5 m; the size of the infiltration pit is 1.8 × 1.8 × 0.4 m.

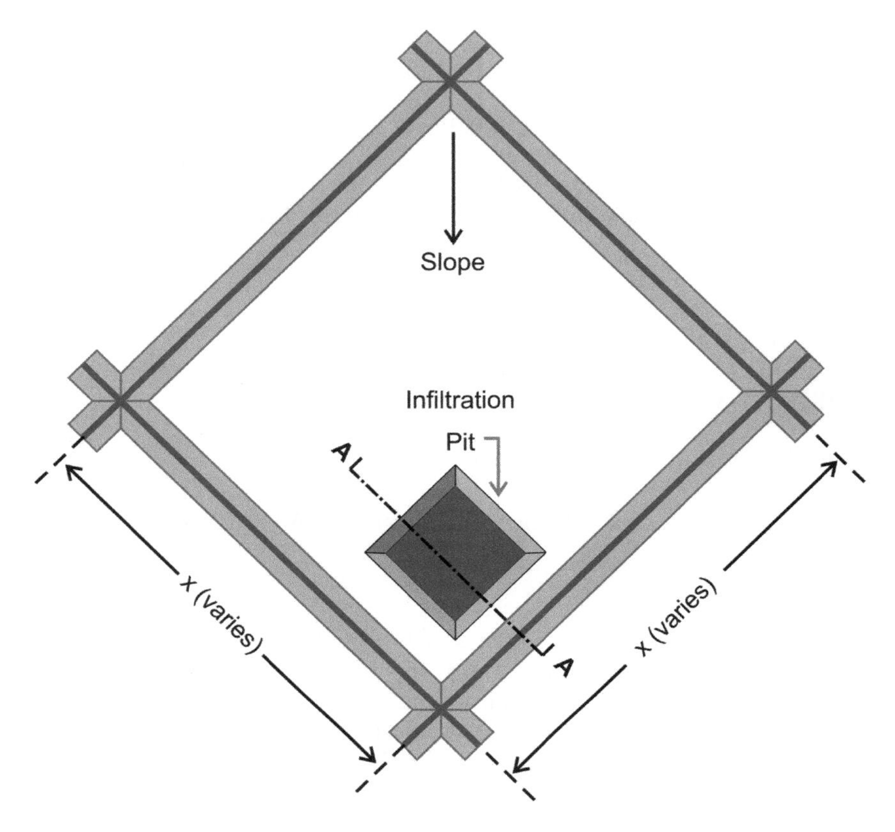

Figure 6.9 Layout of a negarim micro-catchment. (After Crithley & Siegert, 1991).

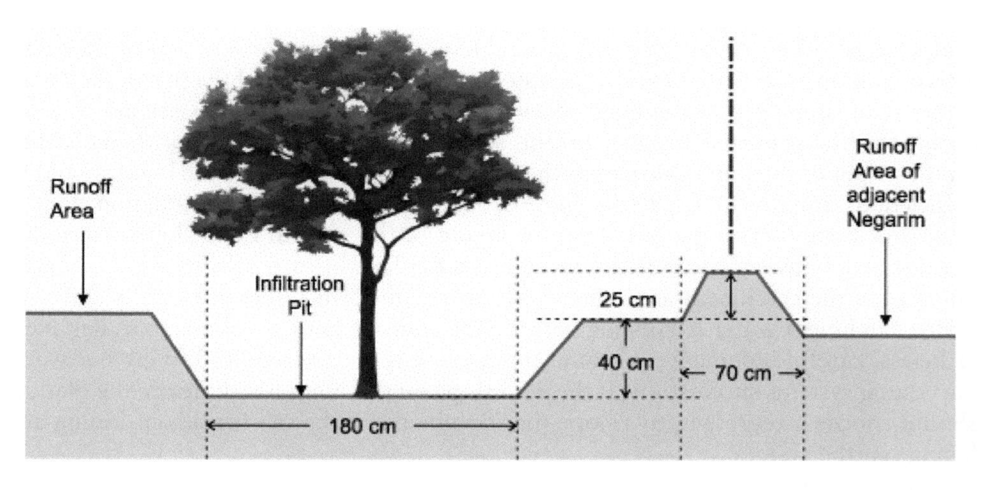

Figure 6.10 Cross-section of a *negarim* (line A – A on Figure 6.9). The tree is planted at the bottom of the infiltration pit. (Prinz, 2010).

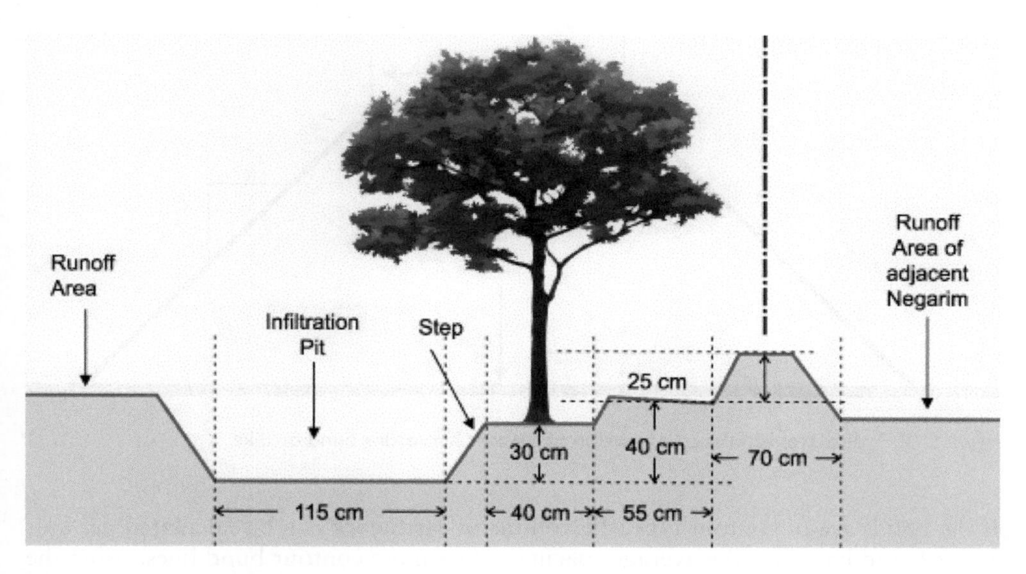

Figure 6.11 Cross-section of a negarim in which the tree is planted on a step if the plant is sensitive to inundation. (Prinz, 2010).

6.5.2 Bund earthwork

Bunds (ridges or dikes) are major components of all water harvesting systems. In this section, the following basic items are briefly discussed: (1) how to estimate the volume of a 1-m length of bund; (2) how to express the bund earthwork volume per hectare; and (3) how to balance the earthwork of a small runoff basin (*negarim*).

Bunds are earthen structures, usually long with a trapezoidal cross section. Figure 6.12 shows a typical cross section of a bund. The volume of earthwork (or stonework) per meter length of the bund can be computed by the following formula:

$$E = d\ (b + z \times d) \tag{6.14}$$

where:

E is the volume of earth/stone work per meter of bund length, m^3
b is the top width of bund, m
d is the height of bund, m
z is the slope of bund sides or faces.

Typical values for z range from 1:1 to 4:1; for b from 0.1 m to 0.40 m; and for d from 0.15 m to 0.60 m.

Thus if b = 0.20 m, d = 0.30 m, and z = 2:1, the volume of a 1-m length of this bund, E, is:

$$E = 0.30\ (0.2 + 2 \times 0.30) = 0.24\ m^3 \text{ per meter length of bund}$$

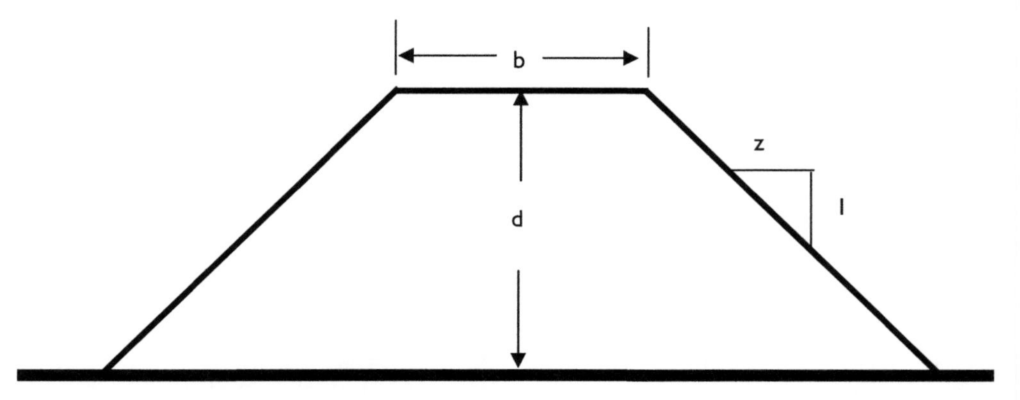

Figure 6.12 Typical trapezoidal cross section of a water harvesting bund or dike.

If the bunds are of contour type, the volume of earthwork can be calculated per unit area of land based on the average spacing between the contour bund lines, using the following equation:

$$V = E \times 10\,000/S \tag{6.15}$$

Using the same values of b, d, and z as in the previous example, if the average spacing, S, between contour bunds was 8 m, then volume of earthwork, V, per hectare of the system land area will be:

$$V = 0.24 \times 10\,000/8 = 300 \text{ m}^3/\text{ha}$$

For small runoff basins (such as *negarims*) with, say, rectangular or diamond shapes, only two sides (one length and one width) should be taken in computing the earthwork per basin. The other two sides are included in the computations of the neighboring basins. For example, if the basins were squares of 5 m × 5 m, b = 0.15 m, d = 0.25 m, and z = 1, then E (from Equation (6.14) will be:

$$E = 0.25\ (0.15 + 1 \times 0.25) = 0.10 \text{ m}^3/\text{m of bund}$$

Thus, the volume of earthwork, V_b, per basin will be:

$$\begin{aligned} V_b &= E \times (\text{sum of length of the two sides}) \\ &= 0.10\ (5 + 5) \\ &= 1.0 \text{ m}^3/\text{basin} \end{aligned}$$

Therefore, the volume of earthwork, V, per hectare of land will be:

$$\begin{aligned} V &= V_b \times 10\,000/\text{area of one basin} \\ &= 1.0 \times 10\,000/(5 \times 5) \\ &= 400 \text{ m}^3/\text{ha} \end{aligned} \tag{6.16}$$

For the semicircular bund, the volume of earthwork per hectare can be estimated from the circle diameter, the spacing between circles' centers along the contour, spacing between contour lines on which the bund tips are located, and the cross section of the bund. For a small semicircular bund on low slopes, the cross section can be assumed constant. However, for large semicircles, the cross section varies. To calculate the volume of a bund or dike with a variable cross section, estimation methods such as the Simpson's or the Trapezoidal rules (Wikipedia, the on-web free encyclopedia) should be used. Alternatively, one can use the average cross section along the entire length of the semicircle. Thus, if the diameter (W in Figure 6.8) of the semicircular bund is 35 m, spacing between contours (Y in Figure 6.8) is 50 m, spacing between circle centers along the same contour is 44 m, and the average cross sectional area of the bund is 0.4 m^2, then the volume of earthwork, Vc, per semicircular unit will be:

$$
\begin{aligned}
V_c &= \text{average cross-section area} \times \text{length of bund} \\
&= 0.4 \times (0.50 \times 35 \times 22/7) \\
&= 22\ m^3 \text{ per semicircular unit}
\end{aligned}
$$

Therefore, the volume of earthwork per hectare of land will be:

$$
\begin{aligned}
V &= 22 \times 10\,000/(44 \times 50) \\
&= 100\ m^3/ha
\end{aligned}
$$

6.5.3 Earthwork balance

In building contour bunds and ridges, earth is usually borrowed from the immediate vicinity of the bund line. The excavated area is immediately uphill (upstream) from

Figure 6.13 Distribution headworks in a macro-catchment scheme in Mali. Channels that deliver water to the cropped fields start from this structure, as does the evacuation channel to get rid of excess water. Structures like this have to be dimensioned as precisely as possible. Photo courtesy W. Klemm/ Karlsruhe University, Germany.

the bund. The excavated area acts like a basin, furrow, or depression for collecting and temporarily storing the harvested water.

For small runoff basins (such as *negarims* for trees), the soil excavated from the infiltration pit is used in constructing the bunds. The amount of soil excavated must equal the volume of earth needed to construct the bunds or ridges of each basin. The standard depth of excavation of the infiltration pit is 0.40 m. Therefore, the dimension of the infiltration pit varies with the dimensions (size) of the runoff basin. In the example for *negarims* given above, the volume of earthwork needed for the bunds for each 5 m × 5 m basin was 1.0 m^3. Thus, the infiltration pit must have a volume of 1 m^3; this can be achieved with dimensions of 1.6 m × 1.6 m × 0.4 m.

As mentioned before, for macro-catchment systems more durable materials and better calculations of runoff and water needs are necessary (Figure 6.13).

Chapter 7

Storage of harvested water

7.1 INTRODUCTION

Water harvesting ensures that a greater percentage of rainfall is put to a beneficial use by concentrating water on productive areas. Water storage conserves surplus water in the rainy season, when rainfall exceeds demand, and allows prolongation of the cropping period into the dry season. The water stored may also cover the demand of crops in dry spells during the rainy season. Generally, there are two means for storing harvested water:

Direct storage in the soil profile, which is usually associated with runoff farming.

Water storage in tanks, cisterns, ponds, or reservoirs.

A storage facility is often required for water distribution supply systems in areas with long dry periods. Where precipitation fluctuates widely over years, storage of harvested water outside the soil matrix becomes an essential part of the water harvesting system.

Factors affecting the selection of the appropriate method of water storage include:

- Topography
- Geology (soil profile characteristics)
- Accessibility for personnel, equipment, and materials
- Costs
- Demand characteristics

Storage requirements must be balanced against the quantity and reliability of precipitation in the area. Storage capacity generally increases as requirements increase and as amounts of available water decrease and become less reliable. Storage requirements can usually be estimated based on the purpose for which the water is to be used, use periods, and water requirement of the intended use (i.e. demand).

All other things being equal, the quantity of water collected from a water harvesting system is proportional to its catchment area. Water losses from storage facilities depend upon the material used for construction of storage system and system type. Storage of harvested water is central to many water harvesting systems. The storage structures should be designed to store enough water to meet demand and constructed in such a way to minimize seepage and evaporative losses. The choice of storage system depends on local hydrological, topographical, and soil conditions.

The best storage system for a given location is the one which supplies enough water to meet demand at the lowest cost per unit of water supplied.

Besides soil profile, the followings are the most common storage mechanisms used for agricultural rainwater harvesting:

- Above ground storage (jars, vessels, prefabricated tanks, etc.)
- Surface/ground storage
- Ponds
- *Hafairs*
- Tanks (India, Sri Lanka)
- Reservoirs (behind dams across *wadis*)
- Subsurface/underground storage
- Cisterns (natural or man-made caves or constructed cisterns made of concrete or bricks)
- Sand-filled reservoirs
- Percolation dams/shallow aquifer artificial recharge
- Underground dams

7.2 SOIL PROFILE

Soil is the natural storage zone of water needed for plant growth. Soil is the water storage facility in all micro-catchment water – harvesting systems and in many macro-catchment systems. Soil, as a storage medium for water, plays an important role in the over-all performance efficiency of the water harvesting system.

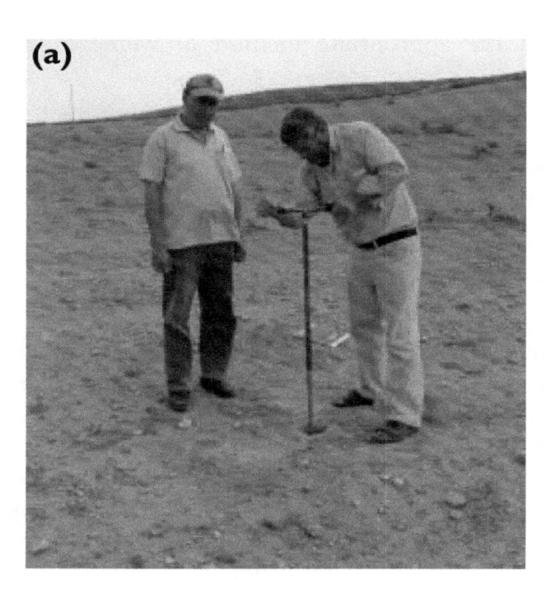

Figure 7.1 (a) and (b): Measuring soil depth and taking soil samples at a potential water harvesting site in Tamasert, Libya. The local soil type is a sandy loam, a soil with a good infiltration and a medium water storage capacity. Photo courtesy D. Prinz, Karlsruhe University, Germany.

Soil, as a porous medium, possesses the ability to retain or hold water within the soil matrix. This retention or storage capacity depends on soil type (texture) and soil structure. Coarse-textured soils are ideal for quick infiltration of precipitation but have low ability to hold water, thus are characterized by small water storage capacity. Generally, these soils are not suitable for water harvesting either as a catchment or as a target component of the system. Medium textured soils (loams) are ideal for the cropped area due to their high water holding capacity (Table 6.1, Box 7.1 and Figure 5.4) and moderate infiltration capacity (Figure 7.1).

The depth of the soil is another important factor influencing water storage. Soil less than 0.5 m will hold too little water to support plant growth during extended dry periods. A depth of 0.5–1.0 m is acceptable; and more than one meter is ideal.

Box 7.1 For how long can a crop cover its water demand from the water stored in the soil matrix?

Assume a **macrocatchment** situation with a healthy, fully grown stand of **sorghum**, growing under subtropical conditions (Figure B7.1.1).

Here are the particulars:

- an effective rooting depth of 100 cm,
- a Reference Crop Evapotranspiration (ETo) of 8 mm/day,
- a Kc factor of 1,
- 100% ground cover, i.e. fully shading the ground, and
- is not shaded by trees etc.

Figure B7.1.1 Sorghum cropping in a long-slope macrocatchment system in Mali. Photo courtesy W. Klemm/Karlsruhe University.

The **soil** is a 100 cm deep **clay loam** with

- 150 mm/m plant available water,
- at full field capacity (i.e. after some heavy rainfall events),
- without compacted layers.

The **weather conditions** are typical for a subtropical region during the rainy season with

- bright sunshine during 8 out of 10 days,
- moderate wind conditions.

Under these conditions the *Sorghum crop will cover its water demand for 2 weeks from the water stored in the soil matrix,* assuming 75% of the plant available water is within the reach of plant roots and no seepage occurs.

7.3 ABOVE GROUND STORAGE

Above ground storage includes jars, vessels, tanks, and containers. Storage tanks can be constructed entirely above the ground, or partly buried. A typical example of the use of jars and storage tanks is in rooftop water harvesting (Thomas & Martinson, 2007; Nissen-Petersen, 2007; Worm & van Hattum, 2006). Rooftop water harvesting is mainly used for drinking water and domestic purposes (see also Chapter 3, Figure 3.6). In agriculture, rooftop water harvesting is limited to gardening and floriculture.

In modern times, greenhouse roofs have been extensively used for water harvesting. Examples from Cyprus and Egypt showed that the water collected from the roof can meet up to 70% of water demand of the crops in the greenhouse (Ben-Asher *et al.,* 1995). Now, with increasing environmental awareness and the desire to conserve valuable natural resources, there is renewed interest in rooftop water harvesting. In many countries where water is a precious commodity, building permits for new constructions are not granted unless rainwater collection storage tanks/cisterns are incorporated into the design.

In many tropical and subtropical countries of the world, rainwater jars or tanks have been used for hundreds of years. In olden times, they were made entirely of clay; now, they are made of various types of cement (Figures 7.2 and 7.3), polyvinyl chloride (PVC, Figure 7.4), burnt bricks (Figure 7.5), corrugated galvanized iron sheets or other materials (Worm & van Hattum, 2006).

Leaving aside the cost of buying or building the tank, the size of the tank is determined by:

- household or garden demand;
- the supply capacity of the roof area; and
- rainfall.

Commercially available tanks vary in size from 500 liters to 10 000 liters or more. The cost per unit volume of water stored in these tanks depends on the construction technique and material used. For large volume storage, two or more tanks can be connected.

To simulate the performance of rainwater harvesting systems using covered water storage tanks, the software program 'SimTanka' was developed. The result of the

Figure 7.2 Jars made of clay (foreground) and modern cement tanks as used in southern Vietnam. Photo courtesy D. Prinz/Karlsruhe University, Germany.

Figure 7.3 Ferro-cement tank as used in coastal areas of Vietnam. Photo courtesy D. Prinz/Karlsruhe University, Germany.

simulation allows the user to design a rainwater harvesting system that will meet demands reliably, that is, i.e. to find the minimum catchment area and the smallest possible storage tank that will meet the demand with probability of up to 95% (For downloading: http://homepage.mac.com/vsvyas/SimTanka2.zip).

Figure 7.4 Pre-fabricated PVC tank for rainwater storage used widely in East Africa. These tanks are commercially available in 1, 2, 5 and 10 m^3 volume. Photo courtesy D. Prinz/Karlsruhe University, Germany.

Figure 7.5 Rooftop rainwater harvesting in northeastern Brazil. The water is stored in half-buried masonry tanks. A manhole allows easy cleaning. Photo courtesy T. Oweis/ICARDA.

7.4 SURFACE/GROUND STORAGE

Surface or ground storage includes all types of storage facilities in which the water body rests on the ground surface while the water surface is open and usually exposed to the atmosphere (i.e. uncovered). Storage structures include excavated pits and ponds, which are easily constructed in relatively flat areas with deep soils. The size of the pond or reservoir is determined by:

- water demand;
- the supply capacity of the catchment;
- the costs of construction and maintenance;
- the potential revenue from the crops produced with the water stored; and
- the finances available to the farmer (Figure 7.6).

To fill these pits and ponds, runoff from house roofs and court yards, from roads (Nissen-Petersen, 2006a), barren lands, rock outcrops (Nissen-Petersen, 2006b) and from pasture areas is collected and directed into these storage facilities. Farm reservoirs are in most cases filled with water diverted from ephemeral streams and rivers.

Figure 7.6 Covered tanks in Ethiopia, constructed with funding from an NGO. This type of tank is excellent in regard to water conservation, but beyond the financial capacity of most small farmers. Photo courtesy T. Oweis/ICARDA.

7.4.1 Small storage ponds

Small storage ponds hold 3000–30 000 m^3 of water. They are normally situated in depressions. Impermeable underlying geology is best for storage ponds.

Ponds in Somalia with a storage volume of about 30,000 cubic meters are made impermeable by compacting the bottom. A concrete inlet reduces the risk of erosion and siltation in case of high runoff.

To reduce percolation, the ponds can be lined with masonry, concrete, or UV resistant, durable plastic sheets (Figure 7.7).

7.4.2 Small farm reservoirs

Small farm reservoirs are very effective in steppe/*badia* environments. To maximize water-use efficiency and reservoir capacity, and to minimize losses in evaporation and seepage, it is advisable that the collected water be applied to the land as soon as possible (if there is demand for water) and stored in the crops' root zone (except for the water needed for drinking and for consumption by animals). Thus the water is best used for the supplemental irrigation of winter crops, during the winter rainfall period, rather than stored for the full irrigation of summer crops.

Figure 7.7 A storage pond in northeastern Brazil, lined with plastic sheets. Photo courtesy T. Oweis/ICARDA.

These reservoirs are usually small but may range in capacity up to 500 000 m^3 (Figure 7.8). Assistance may be needed from an engineer to plan, design and build the dam. The most important feature is to have a spillway with sufficient capacity to allow for peak flows when the reservoir reaches its full capacity (Figure 7.9). Many of the small farm reservoirs constructed in the West Asian *badia* have been washed away through lack of, or insufficient size of, a spillway.

Farmers who have a *wadi* passing through their lands can build a small dam in a suitable location to store some or all of the runoff water in the *wadi* (Figure 7.10) (Oweis *et al.*, 2001).

There are, however problems associated with this technique. The most important of these is the matter of water rights along the *wadi*, which in most of the cases are not well defined. The effect of damming a watercourse on users upstream and downstream of the dam is another issue that causes problems. The best solution for all these problems is to plan the water harvesting intervention within an integrated watershed development approach (Ben Mechlia *et al.*, 2009).

7.4.3 Tanks

The tank system is the backbone of traditional agricultural production in India and Sri Lanka (Agarwal & Narain, 1997). In the early 1980s about 12% of all irrigated farmland in India was under tank irrigation (UNEP, 1983). Tanks can also be used for the cultivation of aquatic vegetables like water-chestnut or even for pisciculture. In China, a new system known as the 'melons-on-a-vine system' uses a combination of

Figure 7.8 Small farm reservoir for supplemental irrigation in Tunisia. Photo courtesy S. Wolfer/ Karlsruhe University, Germany. *(See color plate, page 253).*

Figure 7.9 A small earth dam broken because of the lack of a spillway with sufficient capacity to handle peak overflow. Photo courtesy State Dept. of Western Australia. *(See color plate, page 254).*

Figure 7.10 A masonry dam used to harvest floodwater in a *wadi* in northwestern Egypt. The water is used for supplemental irrigation of field crops and groundwater recharge. Photo courtesy T. Oweis/ICARDA. *(See color plate, page 254).*

tanks and canals for irrigation. In "tank" system, the cropped area is the tank itself. Tanks are above ground water retention earthen structure, constructed on gentile slopes by excavating ground and/or bunding that provide gravity irrigation. A typical example of a cultivated tank or reservoir is the *khadin* system (Figure 7.11). The cost of constructing a tank may be quite low as they are usually constructed manually with few inputs.

Tanks may be below soil surface. In this case, they require some form of lifting mechanism to provide water for irrigation purposes (Sengupta, 1993). Belowground

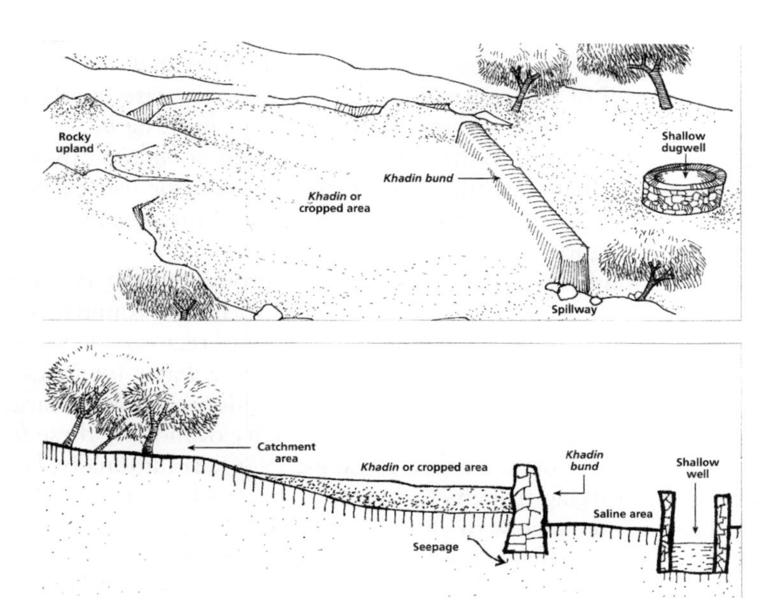

Figure 7.11 Schematic view and cross section of a *khadin* cultivated reservoir (Agarwal & Narain 1997: p. 135).

tanks are sited in such a way as to collect only gentle overland flow with little or no risk of damage by flood. They are also arranged in such a way that water not collected by one is collected by others further down the slope.

Another important water storage system similar to the *khadin* is the *ahar* system developed in Bihar, northern India (Agarwal & Narain, 1997). Unlike tanks, the beds of *ahars* are not dug out. *Ahars* are often built in series so that water can be drained from one higher up the slope to one at a lower level. The bunds of *ahars* follow the contour as much as possible on very gently sloping land. Pipes made of stoneware, concrete, or cast iron with diameters ranging from 150 to 300 mm and spaced some 50 to 100 m apart are used to direct water from the reservoir to the lower agricultural fields. A spillway and a sluice gate should be incorporated for quick removal of water when crops are to be cultivated. It has been observed that brackish groundwater near an *ahar* became potable after the *ahar* was built due to recharge of the groundwater reservoir from the *ahar* (Agarwal & Narain, 1997). The groundwater may also be used for irrigation.

One major limitation of tank systems is the siltation of the cultivated area (the tank), especially in erosion-prone areas. This is becoming more pronounced nowadays because of the scarcity of cheap labor needed to clean and maintain the tank. Management of systems built in series pose even greater problems due to the risk of engineering safety of the earthen structures and inequity of water distribution among upstream-downstream beneficiaries. Tanks can also become a source of disease if they become stagnant, polluted, or infested with insect, snails, and other disease vectors. There is also the risk of people or animals falling into the tanks and drowning. Seepage and evaporative losses are large.

7.4.4 Hafairs

Hafairs are either ponds in natural depressions or excavated tanks with banks (Figure 7.12). The spoil of the excavation may be used to create bunds around the *hafair*. Bunds may be extended upslope to direct runoff into the *hafair* or conduits may be used to direct the runoff to the *hafairs*. The design of *hafairs* can be improved by increasing runoff in their catchment by treating the soil or using a twin system (sedimentation and storage) that facilitates operation and maintenance of the *hafair* (Khouri *et al.*, 1995) (Figure 7.13).

This ancient technique is used to supply water throughout the savannah belt of Africa. In some Arab states, *hafairs* are considered to be the foundation of social stability for rural and Bedouin communities, especially in regions where other water sources are not available. *Hafairs* can be used to provide water for irrigation or to supply drinking water. In Sudan, Kenya, and parts of Ethiopia, however, larger *hafairs* are used to provide water for all purposes. The capacity of most of these *hafairs* lies between 1500 m^3 and 200 000 m^3 (Pacey & Cullis, 1986). Traditionally, *hafairs* were carefully developed, drinking troughs for watering livestock were carefully planned, and rules for their maintenance strongly enforced. In some cases, guards were posted (Sandford, 1983). This type of provision is no longer observed in many regions because circumstances have changed. In Kenya lack of maintenance resulted in almost all the 100 *hafairs* built after 1969 silting up within 10 years.

Being a traditional system, there is the danger that attempts to modernize or modify it may be rejected by the end users. One case is reported from Morocco, where the introduction of modern types was rejected by the local population (Tayaa, 1994).

Figure 7.12 A *hafair* in Jordan. The water in the *hafair* will be available for 3 months after the end of the rainy season. Photo courtesy T. Oweis/ICARDA. *(See color plate, page 255).*

Figure 7.13 An improved *hafair* in Jordan. Animals should not be allowed to enter a *hafair*, but should be watered at a trough outside the basin. Photo courtesy T. Oweis/ICARDA.

7.4.5 Large reservoirs

Dams can be built across *wadis* with ephemeral streams to store water for irrigation. According to construction material, dams can be classified as earthfill, rockfill, or concrete gravity. The selection of type of dam will depend on geology and foundation conditions, topography, spillway size and location, cost-benefit ratio, equipment, and labor availability.

For a narrow *wadi* with an ephemeral stream that flows between high, rocky sidewalls, a concrete gravity dam may be suitable, while level land suggests the use of an earthfill dam (Figure 7.14).

7.5 SUBSURFACE/UNDERGROUND STORAGE

7.5.1 Cisterns

Cisterns are subsurface reservoirs. Their capacity ranges from 10 to 1000 m^3 or more (Ali *et al.*, 2009; Al-Salaymeh *et al.*, 2011). They are commonly used to store water for human and animal consumption, but also for irrigation. In many areas, including Jordan and Syria, they are dug in the rock and are usually small in capacity. In north-west Egypt, farmers dig large cisterns (200–300 m^3) in the earth deposits beneath a layer of solid rock. The rock layer forms the ceiling of the cistern, and the walls are covered with impermeable plaster. Modern concrete cisterns are being constructed where there is no rocky layer (Figure 7.15 and Box 7.2). These may be built using blocks made of cement, burned clay or reinforced cement concrete (Khouri *et al.*, 1995). In Libya, about 7000 cisterns have been installed in recent decades to provide water for animal consumption.

Figure 7.14 A concrete gravity dam in Oman (top) and an earthfill dam in Syria (below). Photo courtesy T. Oweis/ICARDA.

Runoff water is collected from an adjacent catchment or channeled from a remote catchment. The first rainwater runoff of the season is usually diverted away from the cistern to minimize collection of pollutants. Water is usually lifted out of the cistern by a bucket and a rope (Figure 7.16). Cisterns are vital to people in remote areas where no other source of water is available. Settling basins are sometimes constructed to reduce the sediments, but farmers usually clean the cisterns once a year or every other year (Box 7.3).

Figure 7.15 A modern concrete cistern in northwest Egypt. Photo courtesy D. Prinz/Karlsruhe University, Germany.

Figure 7.16 Raising water from a traditional cistern in Egypt. Photo courtesy T. Oweis/ICARDA.

Box 7.2 Cistern water harvesting in the semi-arid mountainous area of Gansu Province, China.

Problem analysis: The water resources per capita in that region are only 230 m^3, while it is 7000 m^3/capita for the world and 2300 m^3/capita in China. Very low standard of living.
Framework conditions: Annual precipitation 330 mm. Unevenly distributed rainfall: 50–70% in July–September, only 19–24% in main crop growth period

- Low runoff coefficient: 0.05–0.08.
- Difficult topography for water conveyance system.
- Groundwater is sparse and of low quality.

Figure B7.2.1 Sedimentation basin and cistern in Gansu Province, PRI China. (Yuanhong & Qiang, 2001). *(See color plate, page 255).*

Figure B7.2.2 'Water cellar' (foreground) in Gansu Province, PR China. (Yuanhong & Qiang, 2001). *(See color plate, page 256).*

Technical solutions: Roads, threshing yards, slopes, and artificial catchments are used to collect and store rainwater in water cellars. Water collected is used to irrigate crops in critical periods to mitigate the effects of drought.

- Water harvesting and irrigation projects are mainly located at villages with favorable catchment conditions, such as close to highways.
- The water cellars usually have a capacity of 30–50 m^3, with a maximum of 100 m^3. A cellar with capacity of 30 m^3 can irrigate 0.13 ha.
- Yield can be raised by 20–40%.

Underground tanks are used to store water to avoid evaporative loss and minimize the cost (Figures B7.2.1 and B7.2.2). The walls of the tanks are covered in cement mortar. The arch-shaped top and bottom of the tank are supported by 10-cm-thick concrete. The cost of this kind of tank is about one-third of that of a tank with thick concrete walls. The total cost of the catchment area and a tank with capacity of 30–50 m^3 is about 1600 Chinese yuan (US$200).

Since the water amount in the water harvesting system is very limited, highly efficient but affordable irrigation methods are used. During sowing, water is applied in the seed hole either manually or via a porcelain pot with small holes in the wall buried near the plants' roots. If the cost is affordable, farmers also use drip irrigation and mini-sprinklers for vegetables in the greenhouse or for fruit trees.

Source: Yuanhong & Qiang (2001).

In the search for low-cost cisterns, a number of alternatives have been tested. Figure 7.17 shows a version developed for small farms in northern Libya. The black plastic sheets, when put in place properly, prevent any light intruding into the cistern. This prevents growth of algae, allowing the water to be used in drip irrigation. There are two plastic sheets lining the pit: the lower one ('Plastic Sheet 1') is left in place permanently, whereas the other one ('Plastic Sheet 2') can sheet lining the pit can be

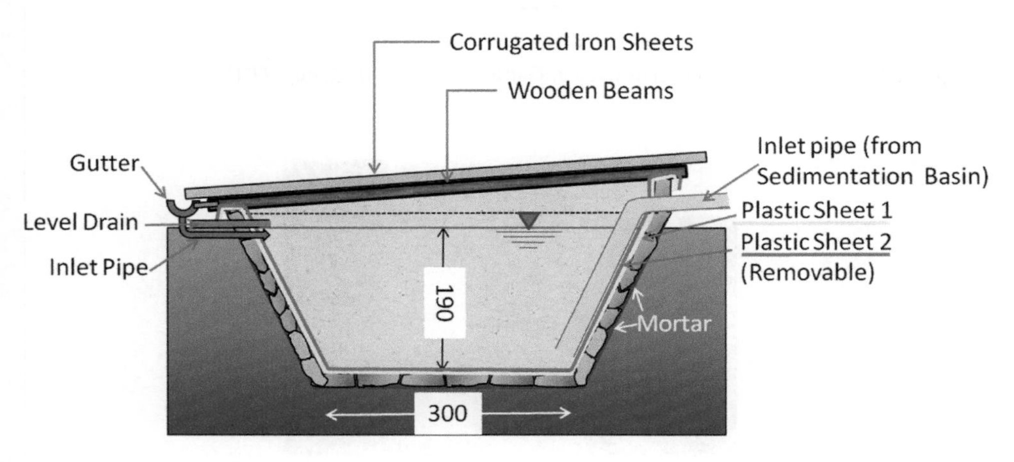

Figure 7.17 Cross section of a low-cost cistern *basin (volume 30 m^3)* developed for rural areas in northern Libya. The basin is covered to corrugated iron sheets; the water running off during a rainfall event is stored in the basin. Source: Prinz (2010).

lifted to remove sediments. A manhole in the roof allows access for cleaning. Rain falling onto the roof is collected and stored in the basin, together with runoff from adjacent areas (Prinz, 2010).

Box 7.3 Case study: Cisterns used in Gansu Province, China.

Problem analysis: 250–450 mm annual precipitation; >70% of rain concentrated in three months; high population density; no alternative water sources.
Goals: Low price/m^3 of water stored, high efficiency in water storage and use.
Framework conditions: The Provincial Government supplies subsidies, farmers contribute labor.
Technical solutions: Rainwater from roads, treated surfaces, and fields is collected in cisterns with a capacity of 30 m^3 (Figure B7.3.1).

Underground storage is used to avoid evaporative loss and reduce the cost. Outside the loess area, concrete cisterns are used, but where soils are loess concrete is used only for the arch-shaped top and bottom.

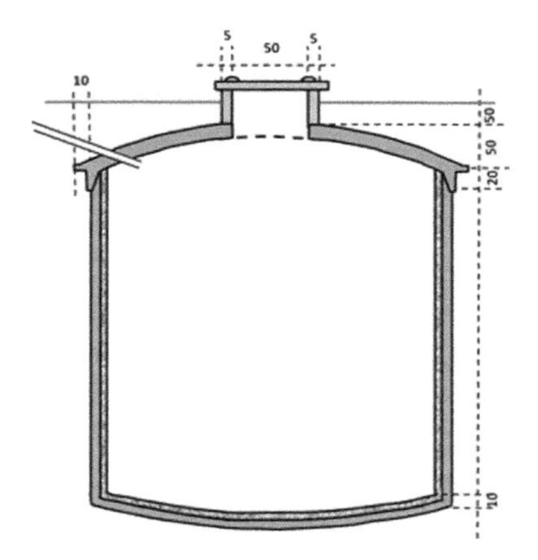

Figure B7.3.1 Design of the cisterns used in Gansu (Yuanhong & Qiang, 2001).

Figure B7.3.2 Water passes through a silt trap before entering the cistern in Gansu Province, PR China. (Yuanhong & Qiang, 2001). *(See color plate, page 256).*

The base of the cistern is a 10-cm-thick concrete slab. The farmers plaster the cistern walls with cement mortar. The cost of this kind of tank is about one-third of that of a tank with concrete walls.

Concrete lined channels direct the runoff water from roads and sealed surfaces to the cisterns. The water passes through settling chambers before it enters the cistern (Figure B7.3.2). The material deposited in the silt trap is usually quite fertile and might be used as fertilizer

Cisterns with capacity of 30 m^3 allow the irrigation of 2 Mu of land (1340 m^2) (Figure B7.3.3). The irrigation methods used are as efficient as possible and affordable to the farmers. An economic analysis is presented in Chapter '9.4.3.3 Examples from China and India'.

Figure B7.3.3 Water from the cisterns is used to irrigate crops. (Yuanhong & Qiang, 2001).

7.5.2 Lining water storage structures

Various materials can be used to line earthen storage structures to minimize losses of water by seepage.

Compacted earth may be used to construct a thick lining. Earth is placed in layers, wetted, and compacted. The layer of compacted earth should be at least 20 cm thick to prevent seepage or percolation.

Sodium bentonite, fine-textured colloidal clay, has been used successfully to reduce seepage in soils containing a high percentage of coarse particles. Bentonite is mixed with soil and applied in a layer 15–20 cm deep, covered with a layer of fine to medium-textured soil.

Membrane and film materials can be used either exposed or buried. When a buried membrane is to be used for seepage control, the earth structure should be excavated to accommodate the cover material. The fill material cover thickness varies from 15 cm to 30 cm with the layer next to the linear not coarser than silt sand. Side slope on which buried membranes are used should not exceed 1:3.

Several types of plastic sheet, including polyvinyl chloride (PVC), polyethylene (PE), and chlorinated polyethylene (CPE), have been used as seepage barriers, mostly buried under the ground. Plastic degrades when exposed to sunlight and is susceptible to mechanical damage. Plastic sheet to be buried should be 25–30 μm think, while that used for an exposed surface should be in the range of 90–120 μm.

7.5.3 Groundwater dams

A groundwater dam obstructs the flow of groundwater and stores water below ground level (Nissen-Petersen, 2000; Nissen-Petersen, 2006c). There are essentially three types of groundwater dams:

1. Sand-storage dams
2. Percolation dams
3. Subsurface dams

7.5.3.1 Sand-storage dams

The sand storage dam is a small dam built on and into the bed of an ephemeral (seasonal) *wadi*. Coarse sand carried by the flow is deposited upstream of the structure while lighter material is carried over the dam (Figure 7.18). The reservoir upstream the dam is thus filled with sand, creating an artificial aquifer that increases in thickness over time (Figure 7.19).

The dam is built during the dry season. It is important that only coarser sediments accumulate above the dam in order to create enough porous space to store water. This can be achieved by constructing the dam from boulders, brushwood, or stone-filled gabions through which fine sand can pass, but that trap coarser particles.

Figure 7.18 A sand-dam in operation. Coarse sediment is deposited upstream of the dam. When the deposit level approaches the top of the dam, the height of the dam may be increased. Consultation of a construction expert is recommended (RAIN, 2007).

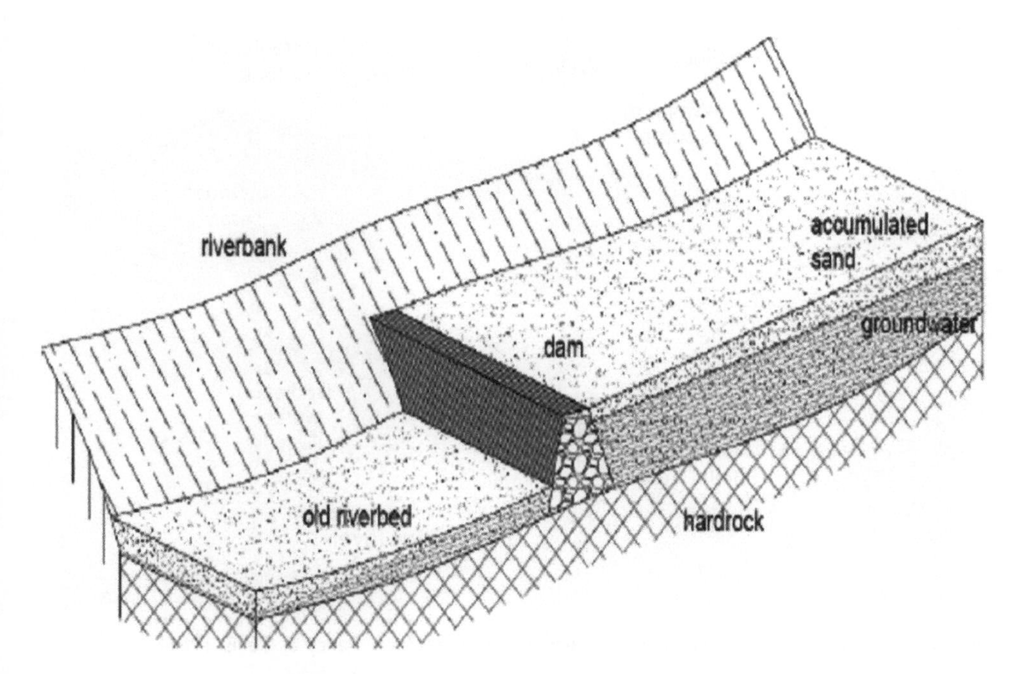

Figure 7.19 Schematic cross section of a typical sand storage dam (RAIN, 2007).

The artificial aquifer is replenished each year and water will be available in the sandy aquifer during the dry season (Hoogmoed, 2007; Borst & Haas, 2006). It takes only one or two heavy rainfall events to fill the sand aquifer completely, and after that the *wadi* will start to flow as it would have done if there were no dam (Quilis *et al.*, 2009). A sand storage dam is most successful if the valley sides below the surface are made of relatively impervious rock, and the bedrock is not too deep beneath the stream bed.

The major advantage of a sand-filled reservoir is that evaporative losses are low relative to an open water surface. Once the water table is more than 60 cm below the surface, evaporative losses cease. Thus, water can be stored for much longer periods than in open storage. Such reservoirs are also at less risk of damage by flash floods. Sand also acts as a filter, and hence the water stored is clean and suitable for livestock, domestic supply, or small-scale irrigation.

7.5.3.2 Percolation dams

Percolation dams are check dams that retain runoff and recharge an aquifer (Figure 7.20). A crop can be irrigated by pumping water from wells. Percolation dams are normally constructed across natural drainage channels. Artificial recharge is used under various climatic conditions to conserve runoff and supplement groundwater resources in shallow aquifers. Several countries around the world have developed national plans for more extensive use of artificial recharge techniques

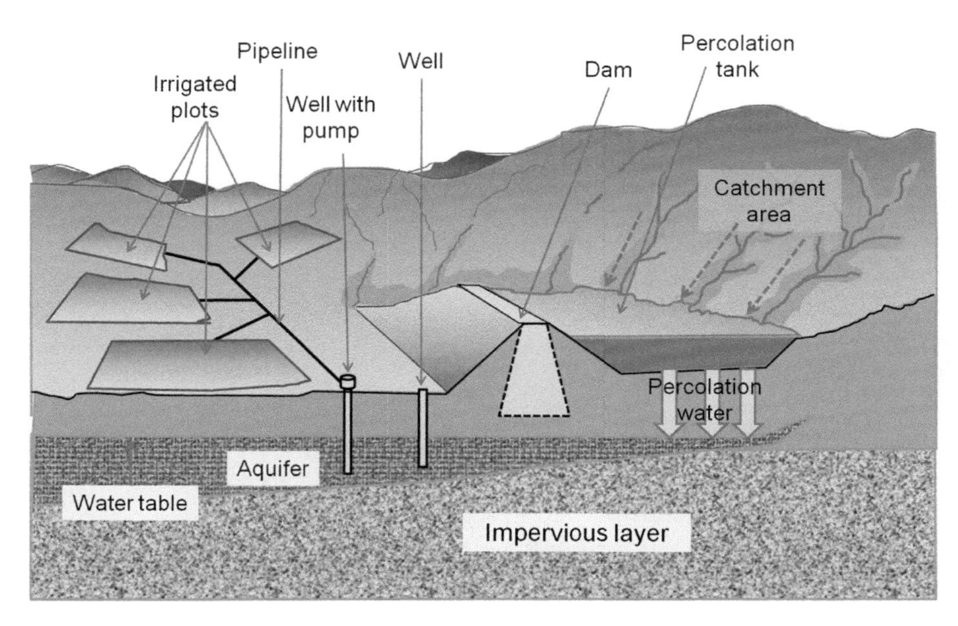

Figure 7.20 Lay-out and cross section of a percolation dam (After Pacey & Cullis, 1986).

(Scanlon *et al.*, 2006; CGWB, 2011). Dams are widely used in Saudi Arabia, United Arab Emirates and Sultanate of Oman and, to a lesser extent, also in Syria and Jordan. In the Arabian region there is a growing recognition of the importance of artificial recharge as a tool for improved groundwater management (Khouri *et al.*, 1995; see Figure 9.1).

Artificial recharge reduces evaporative losses and mitigates the impacts of rainfall variability on agriculture because the water stored in shallow aquifers may be tapped and used as needs arise. Artificial recharge has been used effectively for the storage of harvested flood water in several countries, including Tunisia and Morocco (Khouri *et al.*, 1995). In Egypt artificial recharge was successfully used to increase the amount of water in the Moghra aquifer (Shatta & Attia, 1994: p. 268). A sound geologic survey of the recharge area has to be conducted, if there is the risk of salty layers in the ground.

7.5.3.3 Subsurface dams

A subsurface dam is a vertical barrier constructed in a deep trench such that the top of the dam is at or below ground level (Figures 7.21 and 7.22). The underground dam intercepts and/or arrests the flow in a natural aquifer. As a result, water accumulates behind the dam and the water table rises.

A subsurface dam should have two basic properties: It should be rigid and impervious. It can be used if the valley sides below the surface are of relatively impervious rock, and the bedrock not far below the stream bed.

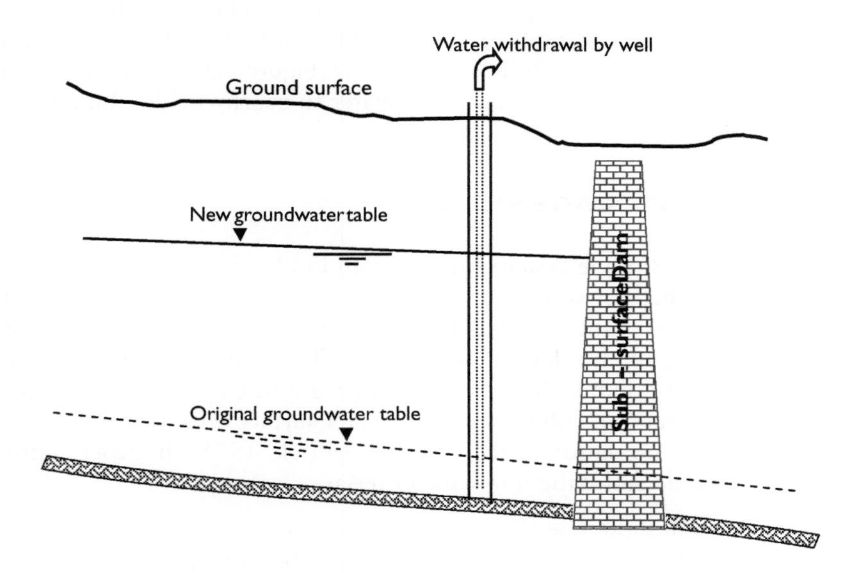

Figure 7.21 Schematic diagram of subsurface dam.

Figure 7.22 Digging a trench in which to build a subsurface dam (RAIN, 2007).

An underground dam can store large quantities of water in a kind of subterranean reservoir in the alluvial fill. For example, a 120-m-long, 2-m-high underground dam in Niger stored enough water to irrigate 110 ha of crops (Figure 7.23). The water can be used by drilling an open well and pumping the water to nearby fields (Raju *et al.*, 2006).

7.6 SELECTION OF STORAGE SYSTEM

It is difficult to design one storage system suited for all kinds of needs and good for all locations. In general, the best storage system for a given location is the one that produces water at the lowest unit cost, has least seepage and evaporative losses, is easy to build and maintain, and is reliable and user-friendly. To minimize seepage and evaporative losses, storage volume should be maximized and exposed surface area should be minimized. The ratio of volume stored to water surface area increases as the depth of storage increases in relation to its breadth (Dedrick, 1975). In general, any water storage system should fulfill the following criteria:

- **Affordable:** The system should be affordable for local farmers. The local economy plays an important role in determining the sustainable benefits from a water harvesting project.
- **Locally available materials:** The material required for building the water-harvesting system should be locally available. Natural materials such as stones and bricks can be used to construct walls, floors, and gutters. Some of the construction material may be recycled, e.g. in India polyethylene sheet made from discarded shopping bags was used to control seepage in small storage tanks.
- **Easy to build:** The water storage structure should be simple and easy to build so that local artisans can learn how to do so within a few weeks. Local knowledge may be applied in building the storage structure.

Figure 7.23 Constructing an underground dam in Niger, West Africa. Photo courtesy Brot fuer die Welt, Germany.

- **Reliable:** The system should be reliable enough to meet demand during the dry season and/or the storage facility should be full at the beginning of the dry season.
- **Easy to maintain:** The system should be easily maintained and repaired using skills available in the local community.
- **Promotes community activities and coherence:** The community should be responsible for maintaining and operating the storage system.

Chapter 8

Implementation, operation, and maintenance of water harvesting systems

8.1 INTRODUCTION

Many water harvesting projects have failed or experienced serious problems (Falkenmark *et al.*, 2001). The main reason for the failures was the lack of an integrated approach during the planning process, especially concerning the implementation, operation, and maintenance of the systems. Lack of maintenance is a common cause of failure of well-intentioned schemes. Projects often assume that the beneficiaries will maintain the system once it is installed, but local communities may not have the resources or skills needed to do this. The cost of maintaining water harvesting structures may be too high for the local population to afford yet may still be low in comparison with the cost of installing a new system. The costs of operating and maintaining the system must thus be taken into consideration during the planning stage so that the system installed will continue to function once the project supporting its installation has ended.

Unfortunately, water harvesting techniques introduced through new projects often ignore traditional methods used locally and introduce new approaches that the beneficiaries are unfamiliar with. Beneficiaries are commonly excluded from any meaningful participation in the planning and implementation process (Bazza & Tayaa, 1994). Even if traditional methods are considered, other systems approaches may be judged to be superior. Where a new system is to be introduced, training in its construction, operation, and maintenance will be needed to ensure the long-term functioning of the system.

This chapter focuses on the operation and maintenance of the various water-harvesting systems discussed in previous chapters. The emphasis is on engineering aspects; socioeconomic issues are covered in Chapter 9.

8.2 IMPLEMENTING WATER HARVESTING SYSTEMS

Beneficiaries should participate in all phases of water harvesting projects, especially implementation. Unless beneficiaries are actively involved in the planning and implementation the project is bound to fail (Critchley & Siegert, 1991; Barrow, 1999). Water harvesting systems should suit beneficiaries' purpose, be accepted by local population, and be sustainable in local environment. The best method applicable in particular environmental and geophysical conditions depends on the kind of crop to be grown and prevalent socioeconomic and cultural factors. Local availability of labor and material are important factors. The accessibility of the site and distance from villages have also to be considered. One of the crucial social aspects for the success is the involvement/participation of the stakeholders or beneficiaries in the planning and implementation of the project. All stakeholders should be involved in planning, designing, and implementation of water harvesting structures. A consensus is necessary for operation and maintenance of these structures. Involvement of governmental and local non-governmental organizations may also enhance collective action by the community (Box 8.1).

Land tenure in rangelands varies from one country to another. In Syria, for example, rangeland is largely public land, but other forms of land tenure such as rented and private land ownership also exist. In Jordan, however, most rangeland is private tribal land. Communal land is commonly overgrazed and little attention is given to sustainability.

Although rainfall is generally greater in mountainous areas than in the rangelands these areas are generally less accessible and home to marginalized and poor communities. The complex landscape consists of steep slopes, terraced croplands, sloping rangelands, and scattered patches of shrubs and trees. Most of the agriculture in these areas depends on direct rainfall. Irrigated agriculture takes place along the banks of the *wadis* that dissect the mountains. The main cause of land degradation here is due to water erosion.

Box 8.1 Water harvesting combined with soil & water conservation in Libya.

Problem analysis: The hilly area is used for fruit tree and barley cropping at a low production level. Substantial amounts of the runoff flow into gullies and consequently most of the area (Hawatim Thella, in NW Libya) is dissected by erosion gullies, which diminish steadily the acreage suitable for agriculture. The erosion gulllies increase in size each rainy season, aggravating the problem. (Figures B8.1.1 and B8.1.2).

Figure B8.1.1 Erosion gully in Hawatim Thella, Libya. The gullies are 6–7 m deep and up to 15 m wide. Photo courtesy D. Prinz/Karlsruhe University, Germany.

Figure B8.1.2 If unchecked, the erosion gullies feed steadily further up-hill, destroying fertile agricultural land such as the fruit tree plantation in the background. Photo courtesy D. Prinz/Karlsruhe University, Germany

Goal: Stabilization of an erosion-prone area by measures which serve simultaneously agricultural production (by water harvesting) and soil & water conservation. The runoff shall be hindered to enter the erosion gullies, but shall be made productive to improve crop yields.

Framework conditions:

- The **soil type** is a silty loam; at some locations surface crusts accelerate the runoff. The **soil depth** in the valley is about 9 m, whereas the soil cover at the top of the hill is merely 0.5 m or less.
- The **inclination** is varying between 2.5 and 10%.
- There are no **rainfall** records of the area available. The **annual rainfall is** estimated to be about 200 mm.
- The **population** density is low; all the land belongs to one clan. Many clan members earn their living outside of the agricultural sector.

Measures to apply:
Construction of

- diversion drains along the uncultivated hilly areas with collection of the runoff in basins and use of the water for supplemental irrigation
- contour bunds or semi-circular bunds in tree plantations to make runoff productive
- inter-row strips for barley production
- infiltration ditches for groundwater recharge.
- protection dams along the erosion gullies to avoid any flow of runoff into the gullies (Figure B.8.1.3).

All measures were discussed in detail with the land owners prior to the start of any activity.

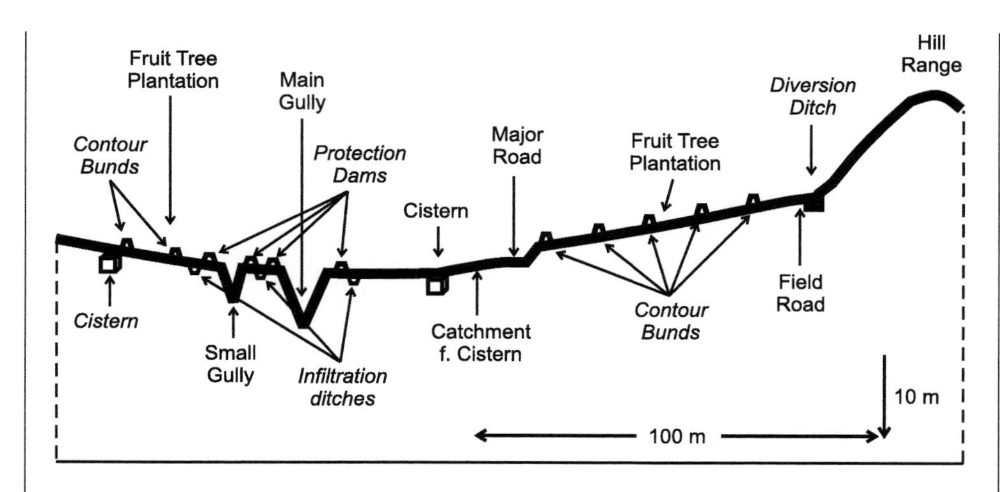

Figure B8.1.3 Cross-section of Hawatim Thella area with water harvesting and soil conservation measures proposed (given in Italics).

Considerable progress has been made in identifying efficient water harvesting and water use schemes for both crop production and combating desertification. Constraints to the implementation and adaptation of these schemes include farmers' unfamiliarity with the technology, conflicts and disputes about water rights, land ownership, and use, and lack of adequate characterization of rainfall, evapotranspiration, and soil properties.

Water harvesting may be implemented by a farmer, the community, or by public agencies. Micro-catchment water harvesting usually comes within individual farms. This is a simple and low-cost approach; however, farmers may experience some difficulty in following the contour lines during construction. Larger micro- or macro-catchment water harvesting systems may need to be implemented through a project authorized by the local community, with help and guidance from the government. Large-scale long-slope and/or floodwater harvesting schemes generally need the intervention of public agencies. Such projects usually involve government facilities or contractors, and machinery or paid local laborers. Here, the initial cost is relatively high. Such so-called top-down approaches are rarely successful; most water harvesting projects implemented by this approach have failed and been abandoned by the beneficiaries.

Projects involving farmers and local communities have been more successful than top-down projects, but require simple demonstrations, training, and extension services. The main advantage of the top-down approach is that it is quick and efficient in rehabilitating land. However, as noted, the costs are high and large systems require expensive repairs that are beyond the resources of local people. It can be justified in areas with high rainfall (i.e. high yield), where little labor is available, and a quick result is needed.

As a first step, all parties involved with the project (farmers, community authority, and government representatives) should be engaged in a round-table-discussion to identify the best technical approaches for the locality. The plan of action developed should be simple enough for the people to implement. Furthermore, the water-harvesting system itself must be sustainable. The planner should be ready to listen and learn from the farmers. Sharing farmers in the managerial role contributes to the success of the project (Figure 8.1).

8.3 CONSIDERATIONS IN IMPLEMENTATION

Many previous water harvesting projects were implemented without a complete integrated study prior to execution, resulting in many technical errors in design and implementation. In many cases, the techniques applied were inappropriate and did not suit local conditions (Figure 8.2). Sometimes appropriate techniques were selected but the installations were neither adequate nor complete. For instance, materials may sometimes not be well compacted, not stabilized with plants or even abandoned half way. Some basic technical criteria that must be met in any water harvesting project (Critchley & Siegert, 1991) are:

Slope: The engineering structures required in a given situation increase as the slope increases. But, water harvesting is not recommended for areas with slopes of more than 50% since this may be uneconomical.

Soils: The soil should be deep, neither saline nor sodic, and should be generally fertile. The major limitations are with limited soil depth and sandy soils which have a relatively high infiltration rate and a low water holding capacity.

Figure 8.1 A researcher of the Institut de Regions Arides, Tunisia, and a farmer in southern Tunisia discuss water harvesting issues. Photo courtesy D. Prinz/Karlsruhe University, Germany.

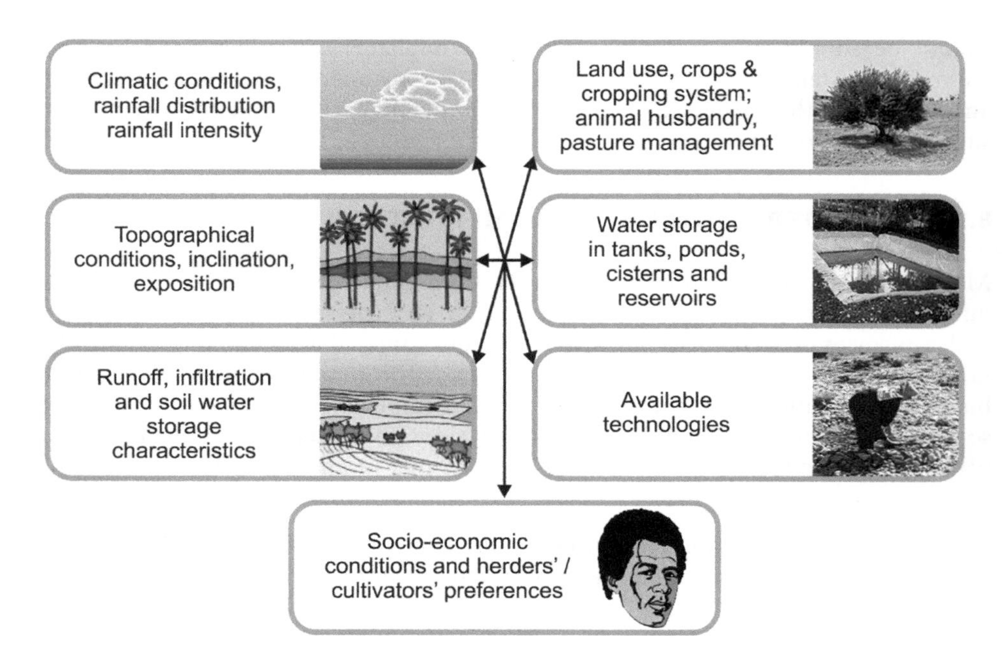

Figure 8.2 Parameters to be considered in implementing water harvesting technique in a specific location.

Costs: The cost of a water harvesting project depends on the amount of earth/stonework involved. Even where this is directly carried out by the farmers, it is a pointer to the amount of labor required to construct the system. Therefore, the selected technique should suit the farmers' financial capability.

In addition to the above basic considerations, the planner must consider alternative sources of water. Each alternative should be compared with water harvesting in terms of cost and level of risk involved. The comparison must take into account not only the water quality required, but also the initial cost and the cost of operation and maintenance (Box 8.2). Where an alternative water source is cheaper to develop, easier to obtain, or can be developed at lesser risk, it should be given priority (Critchley & Siegert, 1991). Other issues must to be considered in any water-harvesting projects to reduce maintenance and overall project cost are given below:

8.3.1 Over-design and under-design issues

Over-designing or under-designing may lead to the failure of the project. Over-design increases project and maintenance costs, and may result in the abandonment of the system soon after implementation. No general rule or guideline can be given on design issues because although water harvesting have been used in some parts of the world for centuries, only very limited design data are available. And even where design data exist, they cannot easily be transferred from one location to the other because of constraints imposed by local conditions.

The impacts of global climate change, such as longer drought periods and higher rainstorm intensities, may require higher catchment: cropping area ratios (CCR) and more solid (i.e. higher, broader and more compacted) earthen structures.

8.3.2 Appropriate technology

Irrigation systems like water harvesting are not just distinct units in themselves. They can be incorporated into existing systems, making these systems more environment-friendly. Several features found in traditional farmer-managed systems can be an important part of the appropriate technology suitable for the locality (Figure 8.3). Some of these systems have withstood the test of time and important lessons can be learnt by studying them. Nevertheless, technology transfer should be carefully planned, especially when a technique alien to the area in question is recommended.

In a government sponsored project carried out in Libya, the planners were aware of the background of the project beneficiaries who have a nomadic way of life and practice primitive subsistence dryland farming (Algharíani, 1993). Most of these nomads lack the basic technical skill necessary to operate and maintain 'modern' farms, which established nonetheless. But no training was provide, although it was apparent to the planners from the beginning that the beneficiaries need some form of training. Moreover, their direct involvement in implementation was delayed until the

Figure 8.3 Using costly machinery as here the 'Dolphin' plough to create large-scale Vallerani water-harvesting basins in the Syrian *badia* is a matter of long-term holistic planning and precise economic evaluation. Photo courtesy T. Oweis/ICARDA.

final stages of project completion. In this example, not only was the intended farming technique alien to the beneficiaries, but also the planners failed to teach and convince them of a new way of solving their problems. These avoidable mistakes contributed in no small measure to the failure of this well intended project.

The many engineering conflicts inherent in water harvesting techniques should not be seen as conflict between the traditional and the modern, but between the sectoral and the holistic approaches; in other words, between ivory-tower expertise and down-to-earth realities. Conflict can be created by the unwillingness of modern technologists to learn from their traditional Counterparts. This conflict should be avoided in water harvesting projects as much as possible. To this end, the technocrats should be ready to work closely with, and accept the opinions of, the farmer beneficiaries.

The success of water harvesting projects depends on good design, economic feasibility and beneficiaries' willingness to undertake full responsibility for operation and management (Box 8.2 and Box 8.3). Operation and maintenance of the system have to be incorporated into the feasibility study, so that the authorities know well in advance what to expect from the project and how its goals will be achieved (Figure 8.4). Good management can only be effective if there is full cooperation between the beneficiaries and the authorities in charge of the project.

8.4 OPERATING WATER HARVESTING SYSTEMS

A locally acceptable team involving mainly the project beneficiaries should be constituted at the beginning of the project to oversee the operation of the water

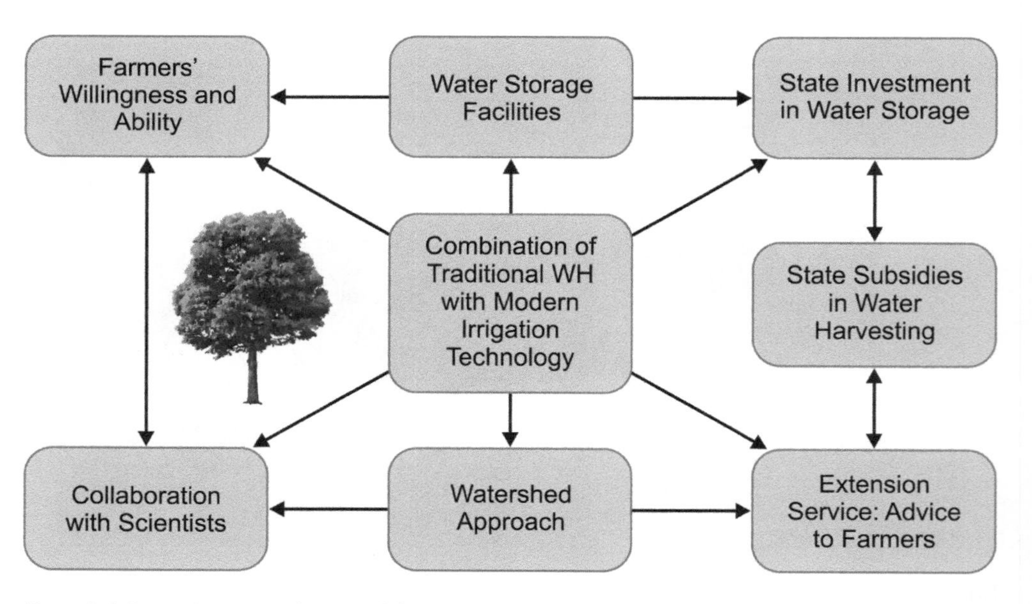

Figure 8.4 Basic elements of successful integration of water harvesting with modern irrigation technology, based on an example from Tunisia.

harvesting system. The responsibility of this body should extend beyond the project phase. The use of an indigenous administrative structure is recommended.

Guidelines and procedures for the operation and maintenance of all components of the water harvesting system should be formulated from the onset of the project. Special attention should be paid to earthen dikes and bunds, water-storage structures, spillways, and diversion structures (Figure 8.5). Micro-catchment water harvesting systems should be inspected after every runoff-producing rainstorm so that any minor breaks in bunds can be promptly repaired.

For a large-scale floodwater harvesting system there may be a need to create a local association that will liaise with the government agency on issues pertaining to the project. This local institution may be supported by the government for a limited period of time, but should be expected to gradually take over the full responsibility of operating and maintaining the project.

All new water harvesting systems should be inspected often especially during the first one or two rainy seasons following construction. Treated catchments should be protected against damage by grazing animals. Silt and trash should be removed from the water conveyance and distribution systems and storage facilities.

Operation of a water harvesting storage facility

In operating a water harvesting system, the user strives to harvest, store, and use as much of the runoff water as possible (Ali *et al.*, 2009). One way to achieve this is to empty the storage facility as soon as possible whenever it gets full. To illustrate this issue, the upper stair-step graph in Figure 8.6 shows the cumulative volume of runoff water coming from a 5-ha catchment at Khanasser area, northern Syria. The total

Figure 8.5 Water harvesting structures such as this measuring weir in a macrocatchment system should be inspected regularly to ensure prompt, effective maintenance. Photo courtesy W. Klemm/Karlsruhe University, Germany.

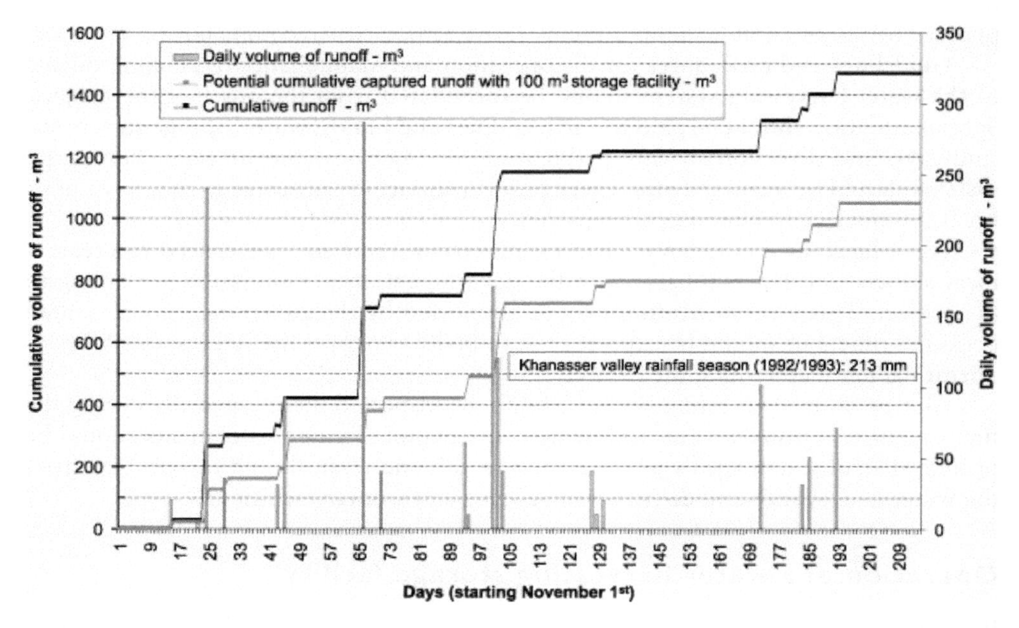

Figure 8.6 Cumulative volume of inflowing and potentially captured and beneficially used using runoff a 100 m³ storage facility at Khanasser area, northern Syria.

seasonal volume of runoff generated from this catchment is 1470 m³. The vertical bars in the Figure represent daily runoff volume. The largest volumes of runoff generated in one day are 290 m³ on day 66, 240 m³ on day 24, 170 m³ on day 101, 120 m³ on day 102, and 100 m³ on day 172. If the capacity of the storage facility is 100 m³, any daily runoff exceeding 100 m³ will spill over, assuming the storage facility is empty at the beginning of each day. Figure 8.6 shows the cumulative volume of runoff water captured and beneficially used in the target area.

Figure 8.6 clearly illustrates that water harvesting depends on only a few intense rainfall events during the rainy season. The total volume of runoff generated by five rainy days amounts to about two-thirds of the total seasonal runoff (920 out of 1470 m³). If the 100 m³ storage facility is to be full at the end of the rainy season, the storage facility should not be emptied after day 180. Seepage and evaporation losses (if any) should be taken into consideration.

8.5 MAINTAINING WATER HARVESTING SYSTEMS

In many water harvesting projects, operation and maintenance are kept under the auspices of a government agency, rather than handed over to the beneficiaries. Maintenance of a water harvesting system requires daily observation to assess effects of heavy rainfall, damage by animals, etc., and this is unlikely to be done if the responsibility lies in the hands of a government agency.

All stakeholders in a water harvesting system should be made aware of the maintenance that is required for the long-term functioning of the system, and must contribute to ensuring that such maintenance is conducted promptly and effectively. Table 8.1 gives an overview of the maintenance requirements of some water harvesting techniques covered in this book.

8.6 MONITORING AND EVALUATION

Water harvesting projects, like many other development projects, are rarely monitored or evaluated to assess how well they are functioning, the extent to which they are meeting design specifications, or the degree and causes of their success or failure. Yet monitoring and evaluation are essential to ensure the effective operation of such projects and to allow operators to take remedial action as necessary. For example, if monitoring shows that the amount of runoff water has been over-estimated, the cropped area can be reduced to achieve the desired results. The information gathered also enables all stakeholders to learn from what went well and what did not, allowing future projects to be modified and improved. Without such information, subsequent projects follow the same pattern as earlier projects, repeating their errors and limitations.

Data should be collected on all aspects of the functioning of the project, including technical performance of the facilities (Figures 8.7 and 8.8), agricultural performance, environmental factors (rainfall, soil erosion, etc.), and socioeconomic and cultural impacts. For large projects, using newest technology like GIS, carefully selected research and monitoring sites, in addition to in-situ ground truth is recommended to enable one obtain an up-to-date information on the project evolution (Figure 8.9).

For orderliness in the collection and dissemination of water harvesting information, governmental agencies can be given the responsibility for regularly collecting, analyzing, and storing data from both ongoing and completed projects (Siegert, 1994). In the already mentioned water harvesting project carried out in Libya, the planners not only forgot to consider the inherent socioeconomic problem of the beneficiaries, but also the issue of monitoring and evaluation before, during and after the project phase (Alghariani, 1994). When the project later collapsed as a result of the many avoidable problems, researchers were called in to conduct some form of monitoring and evaluation using questionnaires. The response of farmers indicated that the issue of land tenure was the major problem that finally rocked the project. This problem could have been detected and avoided if a systematic monitoring and evaluation approach was introduced at the planning stage of the project.

Project planners should ensure that comprehensive reports are kept of all the project phases. These can be used:

- as a means of evaluating the degree of realization of the project objectives;
- to determine the accuracy of the project design assumptions and address any errors;
- as a source of information for planners and government establishments seeking to meet the development need of the region; and
- to evaluate the response of the local populace to the new system.

Table 8.1 An overview of the maintenance requirements and frequency of some water harvesting techniques.

Technique grouping	*Technique name*	*Maintenance requirement*	*Maintenance frequency*
Rooftop water harvesting	Rooftop water harvesting	Cleaning of the storage and collecting surfaces; repair of minor damage	Before the first rain and as soon as damage is noticed
Water harvesting for animal consumption	Traditional techniques	Repair of damaged bunds and removal of debris from the storage/excavations	Before onset of the rainy season and as soon as damage of storage is noticed
	Modern techniques	Repair of the damaged treated surface	Once damage is noticed to avoid contamination
Micro-catchment water harvesting	Inter-row	Immediate repair of any breach on the ridge; catchment area between the ridges to be kept compacted and free of vegetation; ridges to be rebuilt to original height at the end of each season	At the end of each planting season and soon after breaches are noticed; compaction either before or 2 days after the first rain
	Meskat type	Repair of any breach on the manka and the spillway, keep meskat area compacted and free of weeds	Once damage of any type is noticed
	Small runoff basins (*Negarim*)	Repair damage to bunds, remove sediments from infiltration basin, keep runoff area compacted, remove weeds	Soon after damage (or strong basin sedimentation) is noticed, especially after heavy rainfall
	Contour bench terraces	Keeping the bare surfaces free of vegetation while ensuring that excess water drains away freely	No fixed time frame
	Pitting techniques	Removal of debris from the pit after heavy rainfall and redigging of pit after each tillage operation	After every planting season
	Contour bunds and ridges	Repair of ridges in case of overtopping or breaching. Rebuilding of the ridges at the end of the planting season. Ensure a high runoff coefficient	As soon as damage is noticed and at the beginning of new planting season

	Semicircular, trapezoidal and rectangular bunds	Regular inspection and repair of damaged bunds.	Soon after heavy rainstorms especially when the bunds are newly constructed
	Vallerani-type water harvesting	Inspection and technical maintenance of tractor and special plows	Regularly during operation period
Macro-catchment systems	Hillside conduit systems	Regular inspection of the conduits or spillways in areas of higher rainfalls; minor maintenance of the conduits after each rainfall	As soon as damage is noticed and at the end of every planting season
	Tanks and *hafairs*	Removal of debris collected at the base of the below-surface tanks to avoid siltation and repairing damaged section of the bund	Tanks should be cleared of the accumulated debris at the end of every planting season or in between the seasons in case of flooding. Bunds should be repaired once any crack is noticed
	Cisterns	Cleaning silt from the trap and cistern	Once a year
	Limans (tabias)	Inspection and repair of any damaged bund section or the spillway	As soon as damage is noticed
	Stone dams or bunds	Regular observation; replacing dislodged stones in case of heavy rainfall events; plugging small gaps created by runoff with smaller stones; a grass like *Andropogon guyanus* can be planted between the stones to increase the dam height	Specific maintenance should be carried out as soon as damage occurs but general repair should be done after every planting season
	Large semicircular or trapezoidal bunds	Repairing of fissures or breaches in the bund; frequent inspection to find out if rodents burrowed through the bund	Immediate repair of any damaged part; general cleaning and repair annually
	Small farm reservoirs	Frequent inspection of the earth dikes, stone walls, or permeable dams used to block the water	Immediately after damage is observed
	Water spreading	General inspection of the constructed structures after every high rainfall event	Immediate repair of any breach in the structures
	Jessour	Regular inspection of the cross stone wall and the spillway	As soon as damage is noticed

Figure 8.7 A measuring weir in a *wadi* used to monitor runoff in the Syrian steppe. Photo courtesy T. Oweis/ICARDA. (*See color plate, page 257*).

Figure 8.8 Measuring the runoff in a small channel in the Syrian steppe. The small tank in the ground can store about 1 m^3 of runoff. It is advisable to plaster the soil surface around the tank to avoid soil erosion and mistakes in runoff determination. Guiding dikes may be provided on both sides upstream the tank to force all flow to pass over the tank and to prevent erosion on both sides of the tank. Photo courtesy T. Oweis/ICARDA. (*See color plate, page 257*).

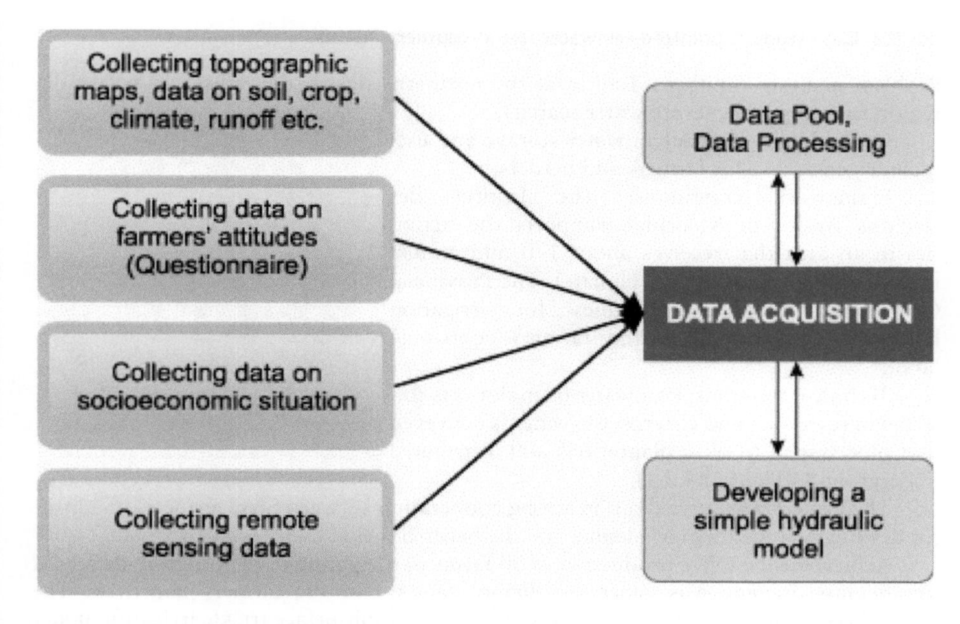

Figure 8.9 Data acquisition for implementing, monitoring and evaluating water harvesting development.

8.7 EXTENSION AND TRAINING

Farmers and extension agents should be trained in the water harvesting techniques employed by the project if the project is to succeed. Training programs should run concurrently with project implementation, using project activities for practical demonstrations. Bringing together project staff, extension agents, farmers, and pastoralists allows the training to address both the role each of them is expected to play in the project and their complementary roles.

Farmers and extension staff can be highly catalytic in the rapid diffusion of simple, low-cost, and efficient water harvesting packages (Reij *et al.*, 1988). This was demonstrated in the Turkana Water Harvesting Project, in which project staff trained local specialists in techniques involved in constructing water harvesting structures. These specialists later trained Turkana farmers interested in improving their sorghum gardens (Reij *et al.*, 1988). The training program is now solely in the hands of the local specialists and the trained farmers are said to be utilizing their knowledge in directing more water into their gardens as it rains.

Similar examples point to the fact that farmers are ready to learn if convinced of the importance of the training. Therefore, each training program should endeavor to identify the most important need of the farmers and articulate means of satisfying those needs. The willingness of the trainer to learn from the farmers' experience will enhance the quality and result of any training program.

Box 8.2 Case study: Optimized rainwater use in southern Tunisia.

Problem analysis: Southern Tunisia at the northern edge of the Sahara is a very dry region suffering from severe water scarcity.

Goals: High efficiency in water storage and use; income generation for farmers and herders.

Framework conditions: The Institut des Regions Arides in Medenine supports the activities in an area that receives about 170 mm annual rainfall (Mediterranean-type climate). The Provincial Government provides subsidies for irrigation hardware and structures, and farmers contribute labor.

Technical solutions: Rainwater from slopes is collected in reservoirs and cisterns; the water is conveyed in a pipe system to olive plantations and distributed underground. (Figure B8.2.1).

The project was carried out in strong cooperation between State agencies, the Institut des Regions Arides in Medenine and the beneficiaries.

Achievements: Olive production of all farms participating in the program increased significantly, income was generated, living standard of the farmers improved. The researchers gathered information on the suitability of subsurface trickle irrigation under farmer's management. (Figure B8.2.2).

Combining water harvesting, water storage and subsurface irrigation results in efficient water use. (Figure B8.2.3)

Figure B8.2.1 The concrete dam catches the runoff of the upstream part of the catchment and a plastic pipeline transfers the stored water to the fields downstream. Photo courtesy D. Prinz/Karlsruhe University, Germany.

Figure B8.2.2 A researcher of the Institut des Regions Arides (IRA) in Medenine, Tunisia, near an opening of the subsurface trickle irrigation system. Photo courtesy D. Prinz/Karlsruhe University, Germany.

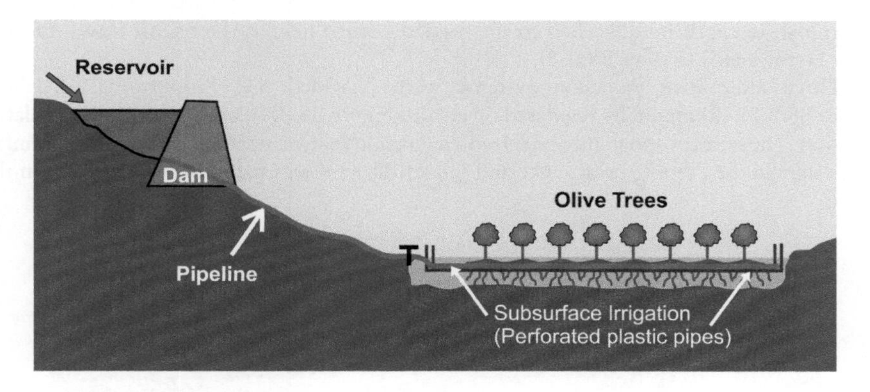

Figure B8.2.3 A cross-section of the macro-catchment project site in southern Tunisia.

Box 8.3 Case study: Macro-catchment water harvesting in western Mali, West Africa (See Box 3.2 for further information).

Problem analysis: Population growth, overgrazing, deforestation, low crop yields, and a high risk of crop failure caused famine and out-migration in the Sahelian region of Yelimane, Kayes Province, Mali (570 mm annual precipitation). The farmers took advantage of natural water harvesting, i.e. they preferred to crop areas inundated after rainfall events, but the growing population forced farmers to cultivate upland sites as well, where risk of crop failure is very high.

Goals: Sustaining life for a rural community under conditions of desertification, decreasing rainfall, and growing population.

Framework conditions: The construction work was carried out by 30 to 50 farmers of the village of Kanguessanou, Kayes Province, under technical supervision and guidance of two foreign engineers (Figure B8.3.1). Financial means were supplied by foreign institutions (European Community, GTZ, Karlsruhe University, Germany).

Technical solutions: Runoff from an 81 ha catchment is directed to a 3.3 ha terraced cropping area and distributed there. The project was carried out according to three basic tenets:

Investments and maintenance costs have to be paid for by the value of the crops harvested.

The management of the system has to be in harmony with the social structure of the beneficiaries.

Equal value has to be given to environmental impacts and economic/management issues.

Preference was given to cultivation of the traditional crops of the region, including sorghum, cowpea, maize, groundnut, rice, and sugarcane.

Farmers were trained and the collecting and distribution system for the cropping area was designed in such a way that the system was flexible enough for changes/improvement, but stable enough to work without skilled personnel.

The conveyance channels collecting the runoff of the slope empty into a distribution basin from there the main distribution canals (and the spillway canal) start. When the uppermost, terraced field is filled to the predetermined height, the runoff flows into the lower laying fields (Figure B8.3.2).

Three alternative diversion systems were provided: (1) Proportional diversion (Figure B8.3.3), (2) manually regulated diversion (Figure B8.3.4), and (3) partially regulated diversion. The farmers apparently preferred the first alternative, not least because most rainfall events start in the evening or at night and one of the sites was half-an-hour's walk from the village.

Figure B8.3.1 Macro-catchment project Kanguessanou, Mali: The construction work was carried out by the farmers themselves; 'Food-for-work' facilitated the activities.

Another aspect was given great importance: The hydraulic structures were designed and constructed in such a way as to work for many years with minimal maintenance, although a certain level of maintenance is indispensable.

Achievements: Sorghum yields could be increased from 0.4 to 1.6 t/ha just by bunding the field (+ spillway) and to 3.3 t/ha by water harvesting (Figure B8.3.5). The risk of crop failure was significantly reduced.

Figure B8.3.2 Flooded fields after a heavy rainfall event.

Figure B8.3.3 Proportional diversion structure.

Figure B8.3.4 Manually-regulated diversion structure. Photo courtesy (4 photos) W. Klemm/ Karlsruhe University, Germany.

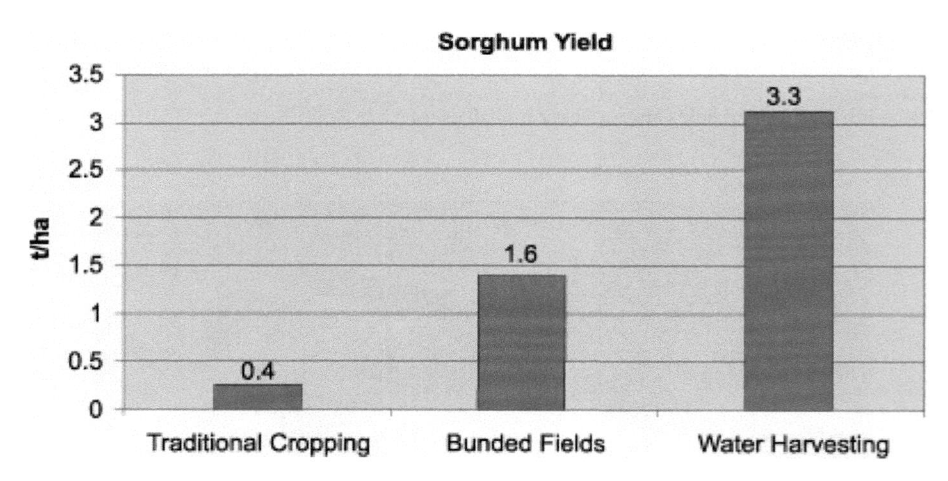

Figure B8.3.5 Bunding and water harvesting dramatically increased sorghum yields in Kanguessanou, Mali. (Klemm, 1990).

Chapter 9

Socioeconomic issues

9.1 INTRODUCTION

Until recently, socio-economic issues received too little attention in many development projects. Many project planners erroneously believe that mere demonstration of new techniques is sufficient to persuade farmers to adopt them (Hogg, 1986). However, the high rate of failure of water harvesting projects suggests the need for more detailed analysis of socio-economic constraints and more dialog with farmers to understand their priorities.

One constraint may be population growth. Population growth in rural areas can impact on water harvesting development in various ways: rising demand for land and water; land fragmentation due to inheritance customs; growth in livestock numbers (often resulting in overgrazing); and increase in poverty incidence, all reduce the capacity and the willingness of people to implement water harvesting.

9.2 SOCIAL FEASIBILITY STUDIES

Water harvesting projects should be based on an evaluation of the long-term gains of the proposed system under existing socioeconomic and environmental circumstances of the project area. This should assess the likely technical, economic, social, and environmental impacts of the proposed water harvesting scheme using a multidisciplinary team. In some cases the social importance attached to the establishment of a water harvesting project may override the commonly derived cost-benefit ratio of many development projects.

Some people argue that social feasibility studies are an integral part of environmental impact assessments (EIAs) and therefore require no separate consideration. However, EIAs tend to rely on quantitative data, while social-impact assessment relies on qualitative data (Barrow, 1987). As a result, social feasibility studies tend to be less precise in their predictions, but even those predictions are valuable forecasting aids.

To make a meaningful social assessment, it is necessary for the assessor(s) to formulate the overall aim of the water harvesting project. Then one can proceed to evaluate the social impacts of the project on the intended beneficiaries. Some of the techniques that can be employed when conducting social feasibility studies include:

(1) check lists; (2) questionnaires, and (3) social surveys. According to Barrow (1987) the commonly assessed social variables are:

- Demographics, e.g. population size, composition, growth rates, mobility
- Human ecological characteristics, e.g. spatial arrangements, housing, neighborhood studies, land-use patterns, accessibility
- Community/institutional arrangements, e.g. amenities and services, community cohesion, community organization
- Cultural aspects, e.g. life styles, world views, beliefs

A useful guide for undertaking social feasibility studies is as follows: Go to the people → live with them → learn from them → love them → serve them → plan with them → start with what they know → and build on what they have (E. Naudascher, personal communication).

9.3 LAND-TENURE ISSUES

Failure to take into account the prevailing land-tenure system in a project area has often hindered water harvesting projects. In the Amamra Water Harvesting Project in Libya (Algharian, 1994), for example, the government developed a strategy to reclaim large areas of the hilly western region of the country and resettle some nomadic sheep breeders. However, the project overlooked land tenure and legal property rights issues. Most of the reclaimed land is the communal property of the local tribes; when such land is divided among families or individual members, Islamic law is supposed to be adhered to strictly. But the project intended to divide the reclaimed land according to proposed farm size, with the result that some of the original landowners were neglected. The plan was to relocate non-beneficiaries outside the project area. However, those affected refused to leave the land as they are entitled to by both Islamic law and tribal custom. This resulted in the sharing of the constructed farms by more than one family and the intensification of disputes among the families and their related tribes.

Land-tenure issues are now problematic in many developing nations. As in many African nations, this became more pronounced after central governments took over land rights and allocated land with little or no regard to the views of the local communities and/or without consulting with them. Ironically, many water harvesting-projects have been sited in communal land because of the large areas involved and their more centralized control, but the absence of statutes defining appropriate land-use rights has sometimes resulted in anarchy (Bazza & Tayaa, 1994).

9.4 ANALYZING COSTS AND BENEFITS OF WATER HARVESTING

Some of the benefits of water harvesting (e.g. contribution to groundwater supplies, soil moisture preservation, soil conservation, control of waterlogging and salinity, flood control etc.) may not be directly measureable and indirect methods (such as a weighted average) must be used to get a reasonable estimate.

In India, detailed meta-analysis of watershed programs has documented the direct benefits of water harvesting, including economic parameters such as benefit: cost ratios and internal rate of returns (Joshi *et al.*, 2009). However, water harvesting in these environments also provides indirect benefits, including substantial environmental and social returns such as combating land degradation, reducing migration from rural to urban areas, and provision of employment. Methodologies for evaluating indirect benefits are sometimes controversial and the private sector is often not interested in these benefits. Economic assessment of macro-catchment water harvesting is more complicated because of upstream–downstream interactions in addition to social and environmental issues.

Even when one attempts to determine the monetary costs associated with water harvesting, the few figures available in the literature cover mainly labor costs. As a result, planners tend to estimate gross benefits of water harvesting in terms of yields and the likely income to be obtained from their sale. From this, they estimate the net benefit of the new technology by subtracting the costs of producing the crop (labor, seed, and other inputs) from the gross benefit. Allowance for risks due to drought, pests etc. should be made by repeating the calculations for good and bad seasons (Pacey & Cullis, 1986).

The following questions should form the basis of a balanced pre-project assessment (Barrow, 1987):

- Is the proposal technically feasible?
- What are the short-term and long-term costs and benefits?
- Who pays and who benefits?
- Are proposals economically viable?
- Do proposals fit into the overall development plan for the region or locale?
- Do the beneficiaries want the project, i.e. is it socially acceptable?
- Is the project flexible enough to adapt to unforeseen changes and altered circumstances?

9.4.1 Costs in water harvesting

The economic success of any new technology depends on the technical effectiveness, social adaptability, and economic efficiency. There are two main types of costs encountered in all production processes including water harvesting: fixed costs and variable costs. Generally these can easily be measured or estimated.

Fixed costs are costs that must necessarily be encountered in the production of the goods irrespective of the quantity of goods to be produced. In water harvesting, fixed costs include workers' wages, cost of machinery, and cost of hardware used in GIS applications.

Variable costs are those that change according to the quantity of goods to be produced. Examples of variable costs in water harvesting include the cost of cultivation of the plot(s), amortization and maintenance costs, costs of seed, fertilizer, contract charges, and soil additives.

9.4.2 Benefits of water harvesting

The use of water harvesting for irrigation purposes has many benefits. The major ones are listed below.

Natural resources conservation: Implementing water-conservation measures, including water harvesting, greatly reduces the risk of flooding, soil erosion, and siltation of water storage facilities in areas prone to such risks. Water harvesting can also contribute to groundwater recharge if harvested water is used to raise the groundwater table (Figure 9.1).

Nutrient harvesting: Water harvesting structures form barriers that retain manure and other organic materials applied to the cropped areas or that are collected from the runoff areas. Thus, nutrient harvesting proceeds at the same time as water harvesting. Such benefits have been reported from locations as diverse as the Sonoran desert (Nabhan, 1984), China, and Burkina Faso.

Reduced costs: In areas with relatively low-input agriculture and scattered settlements, water harvesting offer a cheaper alternative to expensive piped water schemes. This is especially true when the technique implemented is similar to traditional approaches in the region.

Figure 9.1 Recharging groundwater by harvesting floodwater in Saudi Arabia. An assessment of the economic benefits of such a project is hardly feasible. Photo courtesy A.-M. Al Sheikh, Saudi Arabia. (*See color plate, page 258*).

Utilization of local skills: Most water harvesting projects aim at modifying existing traditional farming systems. Local populations in many less-developed nations are capable of constructing many water harvesting schemes, thus leading to the utilization and improvement of local skills. Many other irrigation systems require the use of foreign experts in their construction and operation.

Promotion of self-reliance: In a well-planned water harvesting project, the planners ensure the participation of the beneficiaries from the outset. In this way, the beneficiaries become initiators and managers of the scheme, with the result that they are empowered to duplicate the project elsewhere if need be. Successful projects improve the life of the local populace, thus reducing migration to the cities in search of jobs.

Improvement of the farmers' financial base: For a resource poor rural people, water harvesting generates new direct income resulting from cultivating marginal land and increasing livestock production. However, the net financial benefit to farmers of using water harvesting for plant production can be calculated from differences between net returns from crops produced and weighted average net returns from all rainfed crops. Nevertheless, water harvesting is a gambling on rain business and as such, some sort of probability and risk analysis should be used in its economical analysis.

Other benefits of water harvesting include:

- making grazing and arable land in dry areas productive;
- minimizing risk in drought-prone areas;
- combating desertification by improving pasture growth and permitting tree cultivation;
- supplying drinking water for animals;
- higher crop and livestock yields; and
- offering the possibility of growing higher-value crops.

9.4.3 Economic feasibility analysis

To assess the economic feasibility of a water harvesting technique, specific information on the costs and financial benefits of the system is required. However, this information is seldom found in the literature, thus making it virtually impossible to provide a general cost–benefit analysis of the various water harvesting techniques described in this book. This is particularly the case with traditional systems, which are often neglected by policy-makers.

An economic feasibility study should be carried out jointly by planners and beneficiaries. Ideally, the total costs of the project should be balanced by the overall benefits the project delivers. The initial investment of a water harvesting project may sometimes be a grant, with no expectation of economic return, but even in this case the operating and maintenance costs should be covered by the benefits to make the project attractive to the beneficiaries.

Almost all the economic analyses of water harvesting systems found in the literature are estimates based on the yield of the planted crops and the likely income expected from sales. This 'gross benefit' is compared with the cost of producing the crop, which includes labor, amortization and maintenance, seed, fertilizer inputs, etc. to arrive at the net profit of the new technique (Carruthers, 1983). Maintenance costs are often overlooked in these assessments yet may be much higher in some systems than in others. For instance, the cost of maintaining micro-catchments is relatively high because they need to be inspected after every storm and every minor break in the bunds repaired promptly.

As previously noted, planners, economists, and decision-makers often ignore indirect benefits when conducting feasibility studies of water harvesting projects. These benefits include halting land degradation, combating desertification, supplying drinking water for animals, slowing migration to the cities, minimizing social problems, improving the standard of living of the farmers' families, and enhancing the stability and security of village life. Farmers who implement water harvesting projects in drier environments contribute to these benefits for the general population. However, unless recognition is accorded to some of these special merits of water harvesting, this deserving technology will continue to appear uneconomical. All these aspects must be considered for a comprehensive economic feasibility analysis (Palmier & Nobrega, 2010).

9.4.3.1 Micro-catchments for field crops

The economic feasibility of micro-catchment water harvesting depends on a number of interrelated issues.

First, does the cropped area under water harvesting yield more than the total area (cropped and catchment) under purely rainfed condition, i.e. no water harvesting intervention? For example, if the ratio of the catchment area to the cropped area is 1:1, the yield of the cropped area with water harvesting should be at least double that achieved under rainfed conditions, assuming that the catchment area is cultivable. The rationale behind this question is that there is an opportunity cost for the catchment land, which could be used to grow crops instead of catching water. This is particularly true in the case of micro-catchment water harvesting when the amount of arable land is limited.

Second, how do the fixed and variable costs differ between the water harvesting system and rainfed cropping? Under water harvesting a smaller area is planted with crops, reducing the amount of seed, fertilizer (if any), and labor used.

Third, does the price per unit outputs increase or decrease relative to the cost of inputs? This depends on market dynamics.

By way of an example of how this approach is applied for micro-catchments, Rodriguez *et al.* (1996) conducted an assessment of economic viability of water harvesting for growing wheat and barley in Baluchistan (Pakistan), where total seasonal rainfall ranges between 96 mm and 282 mm. They found that the yields achieved in the cropped area under water harvesting were generally too low to match the production achieved if the whole area (cropped plus catchment) was cropped. However, with a ratio of catchment area to cropped area of 1:1 (i.e. half of the area used for water catchment and half for cropping) net benefits were 23% higher from the water harvesting system than from rainfed cropping and variation in income was reduced by 19% (Figure 9.2). Increasing the ratio of catchment area to cropped area to 2:1 reduced net benefits by 29% relative to rainfed cropping, although variation in net benefits was also lower by 8%. Thus, although water harvesting with a catchment-to-cropped area ratio of 1:1 resulted in lower total output than rainfed cropping it increased both income and stability of income, which are vital to wheat growers in very marginal areas.

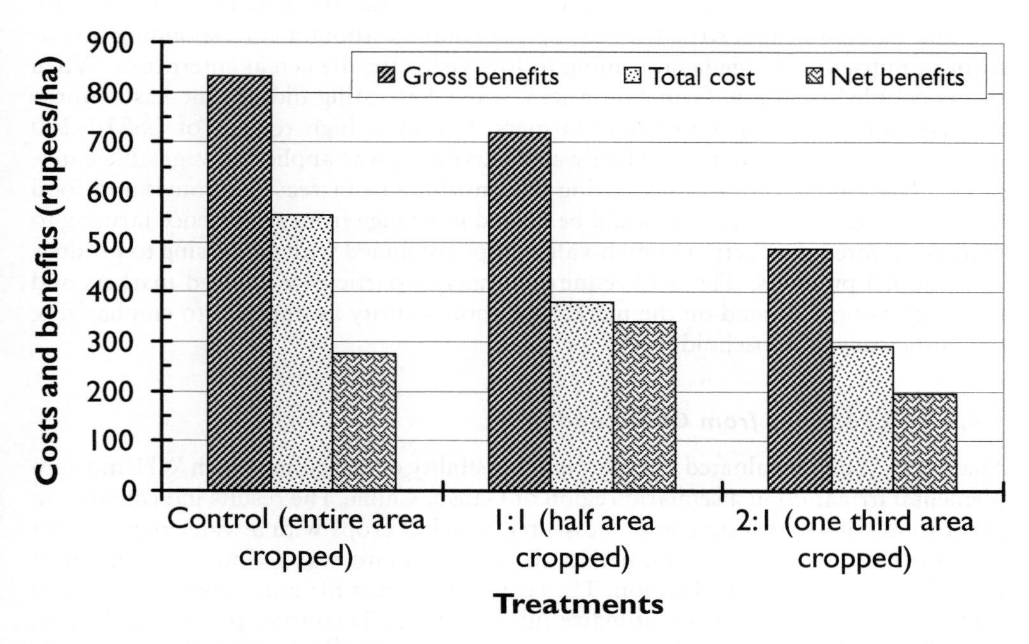

Figure 9.2 Aggregated gross benefits, total costs, and net benefits for wheat grown with different water harvesting treatments in Baluchistan, Pakistan (1986 values, six-year means). (After Rodriguez *et al.*, 1996).

Fooladmand & Sepaskhah (2004) compared various catchment-to-cropping area ratios for grape cultivation under conditions of Bajgah area, Fars Province, Islamic Republic of Iran. Data analyses showed an optimal catchment area of 9 m^2 per vine, yields being 40% larger than under average cropping conditions, i.e. without microcatchments.

Falkenmark *et al.* (2001) stressed the need to supplement water harvesting with soil fertility management and the general improvement of agronomic practices to achieve a net return.

9.4.3.2 *Macro-catchments in sub-Saharan Africa*

Water Harvesting is being promoted widely as a way to increase the production of crops and livestock in semiarid areas of eastern and southern Africa (ACPC, 2011; Cullis & Pacey, 1992; Hogg, 1986; Pachpute *et al.*, 2009; Reij *et al.*, 1988; van Dijk & Ahmed, 1993; van Dijk & Reij, 1994). In Tanzania, farmers are using water harvesting technology to produce maize, paddy, and vegetables in semiarid areas where it would otherwise be impossible or very difficult to produce these crops (Hatibu *et al.*, 2006; Mzirai & Tumbo, 2010). The economics of these practices has been assessed in two contrasting districts over a period of five years. Gross margin analyses were used to assess economic performance of various water harvesting systems with respect to return to labor and income generation. The systems evaluated included macro-catchment with floodwater diversion and small storage ponds.

Results show that most farmers have invested large amounts of labor to establish and maintain earth structures to capture runoff without a corresponding investment in nutrient management, leading to low yields for the cereal enterprises. When this is coupled with low farm-gate prices, water harvesting did not increase returns to labor for the majority of cereal farmers. However, high returns of US$10–200 per person-day were obtained when water harvesting was applied to vegetable enterprises. Therefore, for water harvesting to contribute to increased incomes and food security, smallholder farmers should be helped to change from subsistence farming to market-oriented production of high-value crops combined with processing to produce value-added products. This will require farmers to participate in food markets and thus increasingly depend on the market for food security as opposed to emphasizing self-sufficiency at household level.

9.4.3.3 Examples from China and India

Yuan *et al.* (2003) evaluated the economic feasibility of agriculture with WH and supplemental irrigation in a semiarid region of Gansu, China. The results indicate that in order to maximize investment, it is essential to select crops with a water requirement process that coincides with local rainfall events. Potato was found to be the most suitable crop in the studied region. The economic indices for potato were superior to spring wheat, maize and wheat/maize intercropping. Therefore, potato production using WH and supplemental irrigation is the best alternative for cropping systems in the semiarid region of Gansu, China.

For resource poor smallholder farmers in water scarce areas, even small volumes of stored water for supplemental irrigation can significantly improve household economy. In the Gansu Province in China, small (10 to 60 m^3; on average 30 m^3) subsurface storage tanks are promoted at large scale (Qiang *et al.*, 2007). (See Box 7.3 for further details.) These tanks collect surface runoff from small, often treated catchments (e.g. with asphalt or concrete). Research using these subsurface tanks for supplemental irrigation of wheat in several counties in Gansu Province (Li *et al.*, 2000) indicate a 20% increase in water use efficiency (rain amounting to 420 mm + supplemental irrigation ranging from 35 to 105 mm). Water use efficiency increased on an average from 8.7 kg/mm/ha for rainfed wheat to 10.3 kg/mm/ha for wheat receiving supplemental irrigation. Similar results were observed in maize, with yield increases of 20 to 88%, and incremental water use efficiencies ranging from 15 to 62 kg/mm/ha of supplemental irrigation (Li *et al.*, 2000).

More economic data are available in India which practices flood water harvesting or what is locally known as 'tank irrigation'. According to the 1975 survey of research team from ICRISAT (von Oppen & Subba Rao, 1980), a hectare of land under tank irrigation produced about 300% increase in food grains in comparison to an unirrigated hectare. In carrying out this work, the cost of constructing old tanks was difficult to ascertain by the researchers leading to their using a proxy. By including the estimated cost of tank construction, project life of twenty years, and a discount rate of 10 percent; it was found that only 28% of the total tanks survey has a benefit-cost-ratio (BCR) of greater than unity and less than 2% had an internal rates of return (IRR) above 5 percent (Sengupta, 1993). This is the general economic picture of water harvesting systems in India. If only the maintenance and operational costs are considered, some of the tanks are economically viable since productivity can be increased. But if construction costs are included, some of them may be uneconomical (Falkenmark *et al.,* 2001).

The storage volume might be an important parameter, too: When comparing on-farm rainwater reservoirs with centralized large reservoirs, studies conducted in Kharagpur, West Bengal, India, showed a gross return of US$ 0.15 per m^3 of water per year for small and US$ 0.05 for large reservoirs (Pandey *et al.,* 2011; van der Zaag & Gupta, 2008).

9.4.3.4 Some general recommendations

To make water harvesting systems a promising alternative to all involved with the promotion and utilization of the various techniques, there is need to make suitable alterations in the methods of cost-benefit calculation. Planners, economists, and decision-makers often ignore indirect benefits when conducting feasibility studies of water harvesting projects, it is essential to understand the importance of these too. They include halting land degradation, combating desertification, supplying drinking water for animals, slowing migration to the cities, minimizing social problems, improving the standard of living of the farmers' families and enhancing the stability and security of village life. Farmers who implement water harvesting projects in the drier environments will be contributing to these benefits for the general population. Unless recognition is accorded to some of these special merits of water harvesting, this deserving technological alternative may appear uneconomical. The following modification may be necessary (Sengupta, 1993; Oweis *et al.,* 2004a):

- Include all kinds of benefits arising from the system which are neglected in the current method of calculating Benefit Cost Ratio, BCR, (e.g., social and environmental benefits). Many of the water harvesting systems contribute substantially through groundwater augmentation, facilitation of water availability for animals, preservation of soil moisture etc.
- Cost and benefit due to environmental impact like flood control, reduction in waterlogged area, soil conservation should also be considered. No standard environmental impact assessment is yet possible nor are such qualitative benefits like survival value in the face of severe currently quantifiable.

All these aspects must necessarily be considered for a comprehensive economic feasibility analysis.

9.5 INTEGRATED APPROACH TO PLANNING AND MANAGEMENT

Publications on water harvesting present more information on technical specifications than on planning and management. Reports on water harvesting projects in the literature tend to focus on projects implemented by government or donor agencies rather than on those undertaken by individual farmers or communities (Bazza & Tayaa, 1994; Oweis *et al.,* 2004b).

Nevertheless, a prerequisite to good management is that those concerned with the project know what they expect to achieve from the project and how to go about it. Furthermore, the beneficiaries have to be engaged with the project from the outset; they should agree with the objectives of the project and their expected role in it. The managers of water harvesting schemes should endeavor to understand the way of life of the beneficiaries. In many parts of the world one can see changes in the collective way of life that existed in the past and provided for people's livelihoods and survival. This system of living has progressively been shifting towards private ownership, individualism, and personal interests. Therefore, one should always try to negotiate new and acceptable terms in case of collective ownership of projects. This is a call to an integrated approach to the planning and management of water harvesting projects.

An integrated approach to the planning and management of water harvesting projects will lead to the development of a system that will sustainably utilize the harvested water to improve agricultural production while being socially and economically beneficial to those utilizing it. In this approach, one does not concentrate on finding the best water harvesting technique or on analyzing at what level management should concentrate. Rather the focus is on recognizing that planning and management are integral functions of all those involved in any water harvesting project. Figure 9.3 gives an overview over the parties which are involved or can contribute to a successful implementation of water harvesting for agricultural purposes:

- Parties involved in water harvesting planning and management
- Parties involved in fixing the frame conditions

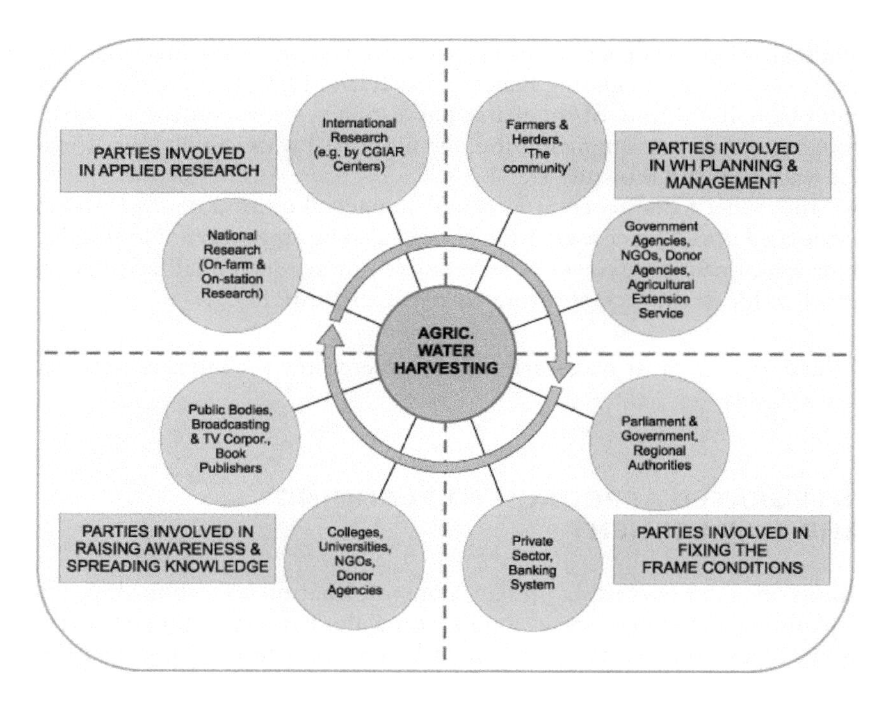

Figure 9.3 Parties involved in water harvesting for agriculture.

- Parties involved in raising awareness and spreading knowledge, and
- Parties involved in applied research.

The various roles expected of the parties involved should be categorized, with emphasis on the respective contributions of each to ensuring a successful water harvesting scheme. Given below are some of the roles expected of government agencies, the community, farmers, and donor agencies in a water harvesting project. The functions of each as stated here are by no means exhaustive and may vary from one area to another (FAO, 1997).

9.5.1 The role of government agencies

In many water harvesting projects, government agencies are expected to play important roles that will generally be for the benefit of all parties involved in the project. The most appropriate approach is to ensure that consensus of all parties is achieved before any regulation or policy regarding the project is enacted.

Common roles for government agencies involved in water harvesting projects include the following:

- Act, at the request of the beneficiaries, as the middleman between donor agencies and the beneficiaries.
- Enact appropriate regulations, in consultation with the beneficiaries, in support of the goals, functions, and financial obligations of the parties involved, the expected targets, etc. In doing so, the authority concerned should specify the rights and obligations of all parties involved in the project.
- Attempt to create social groups at the local level that will be responsible for the day-to-day management of the project. The ultimate aim of this is to have the local groups assume direct responsibility for the management of the whole project. This might require a long period of negotiation with key individuals and institutions within the target area.

Farmers should be convinced that government is only acting on their directives and as such their input is important. Government agencies should avoid imposing their views on the farmers because doing so has led to the abandonment of many projects soon after completion.

Policies concerning the operation and maintenance of the system should also be drawn from the outset of the project. This may require the creation of a local association that will liaise with the government agency on issues pertaining to the project. This local institution may be supported by the government for a limited time, but should be expected to become an autonomous local institution with government approval (FAO, 1997).

9.5.2 Community participation

Only recently has the issue of community participation become an important consideration for project planners, who initially considered that a mere demonstration of new techniques sufficient to persuade farmers to adopt them. Very little has actually been written on this issue with respect to water harvesting, but experience has shown that unless people are actively involved in projects aimed at improving their standard of living, the project is bound to fail. However, it is important to note that wooing a whole community to participate in a water harvesting project may take a long time.

Project planners often wish to impose their ideas, priorities, or even systems on a community. This approach has often been met with strong resistance from the local populace whose views and expertise are overlooked. If water harvesting projects are to be safeguarded against the danger of being regarded as merely a reflection of technological enthusiasm, ordinary people must be involved in their planning and implementation. Imposition of techniques and modalities of implementation (top-down approach) have led to the failure of many soil and water conservation projects in sub-Saharan Africa (Reij *et al.,* 1988).

The involvement of all parties in a water harvesting project calls for innovative dialogue and partnership (Pacey & Cullis, 1986). The best beginning point of this dialogue is the assessment of the priorities of the local population before deciding on the type of project needed by them. This assessment may show that the local populace has a different need than that identified by the outsiders; in this case, theirs must be considered if betterment of living standard is the aim of the funding agency.

On a broader note, the moment an outside government or funding agency representative arrives in a given area, he/she should spend some time in the area studying the lifestyle of the people and the various activities from which they earn their living (see 'Social feasibility studies', above). This will also involve finding out the administrative structure of the community and some form of integration into the pattern of life. During this period of orientation, one is likely to find out if individual or collective water harvesting projects will be best for the region in question. In many African countries, it seems that farmers favor an individual (family) approach to water harvesting rather than a collective approach (Reij *et al.,* 1988). This may, however, be an exception rather than the rule. Each area should be considered individually.

In an analysis of changes that occurred in a local water harvesting technique in Mexico (Doolittle, 1984), it was discovered that structural changes in the cultivation of floodwater fields took place gradually as a result of improvement of the financial circumstances of individual farmers. The development of the system continued on a somewhat trial-and-error basis until the system changed from individual resource management to communal management.

Experience has also shown that voluntary participation in development projects is rare, and as a result project planners often resort to the use of incentives like cash, food-for-work, etc. The danger of using such incentives is that farmers will just see themselves as laborers rather than participants in a development project. This might lead to the demise of the project soon after completion since beneficiaries will often be incapable of maintaining the structures or have no interest in doing so. Hogg (1986) found that the use of food-for-work in a water harvesting project in Kenya was one of the major constraints to its successful implementation. However, Reij *et al.* (1988) suggested that food-for-work should only be used in years of food shortage in a communal project where benefits do not accrue directly to the individual farmers. This should, however, stop as soon as normalcy returns to the locality.

9.5.3 Gender representation

In many developing nations the issue of who does what is very important for the local populace. Most of the labor input in agriculture comes from women, but some functions are entirely reserved for men. The planners of a water harvesting project should be aware of the potential roles of men and women in such projects. Where little or no such formal demarcation exists, experience has shown that women play more important roles in water harvesting projects than men. In a water harvesting project in Kenya, for example, the women employed and trained for the project were willing to share their knowledge with other women, whereas the men were not (Cullis, 1987).

Ideally, more women than men should be empowered to manage a water harvesting project, where restrictions do not exist. This is because women provide the major share of agricultural labor in many developing nations; men are likely to seek off-farm employment after the harvesting season. Even in communal projects, women more easily organize themselves than men and are better at holding onto their convictions once informed of the importance of the project to the betterment of their standard of living.

Nevertheless, project planners should ensure that the project does not improve the standard of living of one gender at the expense of the other. One gender may have more roles to play in the project than the other, but the overall effect should be to the benefit of all the local people.

9.5.4 Farmers as managers

The people affected by a project have the right to be democratically involved in all deliberations concerning the project from the beginning (see 'Community participation', above). For projects like water harvesting which aims at promoting sustainable development, this should be self-evident (Naudascher, 1996).

Farmers should be invited to participate in water harvesting projects to share their expertise. One can rightly argue that not all farmers have the same level of competence in local farming practices, but the researcher or project planner must be ready to treat each farmer as an expert in his or her field. Neglecting local expertise has resulted, in many cases, in the imposition of alien farming systems and techniques on the local populace; such approaches are not sustainable. Therefore, a good practice is for project staff to make clear from the beginning their intention to learn from local farmers (Werner, 1993). To minimize the farmers' suspicion of outsiders, the project planner should make the farmer feel an important partner by, for example, asking him or her to explain some of the local farming techniques or practices relevant to the proposed project. The planner, by being ready to listen and learn from the farmers, will communicate the message that local farming practices are worthy of respect, which is important in cultures where farming is regarded as a low-status employment.

When farmers are treated as managers of a water harvesting project, planners will be gradually taught the customs of the community while enjoying the hospitality and cooperation of the local people. This also involves showing respect for the farmers' time by finding out how much time the farmer has for the proposed activities. Placing the farmer in a managerial role promotes dialogue on innovation, because it provides the planner with local agricultural terminology, which is very important in understanding the farmer's concepts. It is also an opportunity for the planner to assess how articulate each farmer is, as they explain why and how local practices are used (Werner, 1993). This will be useful when allocating different tasks to different farmers as the project progresses.

The Lokitaung Pastoral Development Project in Turkana, Kenya, is an example of a successful project in which local people had a considerable degree of control over what was done in the project (Cullis & Pacey, 1992). The project not only used locally adaptable water harvesting technology, but decisions were based on priorities identified by the local people and development workers were ready to allow the work and its goals to evolve over several years, resisting external pressures on meeting target dates.

9.5.5 The role of experts and donor agencies

Experts involved in a water harvesting project must be ready to commit time to learning from the community where the project is to be sited, giving time for personal relationships to grow and learning experience to be digested (Cullis & Pacey, 1992). Experts should also be aware of the different kinds of knowledge characteristic of educated, urban people and rural people with less formal education. The emphasis of many experts, especially those with a western-type education, is to discover basic principles of a particular system that can be universally applicable without considering local conditions. However, in water harvesting projects they should pay more attention to local knowledge of agriculture and water harvesting techniques and be ready to adopt or adapt these as need be.

Furthermore, the experts, in constant consultation with the local people, should come up with a workable plan for how to institutionalize the management of the project. This will ensure the continuity of the project under the control of the beneficiaries. A research unit responsible for disseminating information about the project should be included within the framework of an institutionalized setting, to enable an easier duplication of the project techniques to a relatively similar location if and when necessary.

Many development projects have failed because of the pressure on the part of the sponsors for quick and tangible results within a specified time limit. If water harvesting projects in developing nations are to succeed, donors should begin to think in terms of processes rather than projects. To differentiate between the two, a project is limited by time and predetermined targets and the success or failure of the project is based on hitting those targets within that time limit; a process goes on indefinitely, with interlocking phases of success or failure and consequent improvement in subsequent phases.

Donor agencies should be ready to make longer-term commitments once convinced of the value of a water harvesting project. They also have a role in ensuring that funds for the project are used for the stipulated purposes. Experience in many developing countries show that governments divert funds meant for development projects to other areas, especially where the government's priorities conflict with those of the donor agencies. Donor agencies should ensure that this does not become the bane of water harvesting projects.

On the other hand, when a water harvesting project has received the approval of a donor agency, guidelines for its implementation should not be too rigid, so as to enable development workers and beneficiaries to make adjustments within their laid-down conditions without jeopardizing the overall aim of the project. If donor agencies can become less demanding with respect to the achievement of defined goals within a specified time frame, water harvesting projects can contribute greatly to meeting the water needs of farmers in many dry areas of the world.

9.5.6 Adoption or non-adoption of interventions

All parties involved in a water harvesting project must decide at the beginning the ultimate aim of the project. If the project is aimed at the improvement of the farming activities of the rural dwellers, then they must play a leading role in all the phases of the project until the management is entirely handed over to them. Even the techniques finally chosen by them must be capable of being operated and maintained with local resources.

In his evaluation of the Turkana water harvesting scheme (Kumu, 1986) pointed out that only 15% of the dryland farmers in the Baringo region of Kenya were found to be practicing water harvesting four years after the first experiment with water harvesting started in that region. Reasons given for the low rate of adoption of the water harvesting techniques included:

- The belief of some local farmers that the initial labor input in the newly introduced system was higher than their traditional technique of deep tillage
- Bad design of initial structures
- Poor approach of the extension workers who introduced the new techniques to the farmers.

These and other problems identified all point to the crucial importance of management in any water harvesting project. When the local people and their institutions are active participants in a project, their views and suggestions can shape the project to meet their needs, increasing the likelihood of uptake of the project's interventions.

Many failed projects in developing nations were conceived on the concept of 'technology transfer' in which 'experts' (usually from donor countries) come in with ideas of what they think will usher in needed development in the area in question. By implication, these experts assume that the local people have nothing to offer but should just copy and implement what the experts bring.

This is a recipe for failure. No matter good a planned project may look on paper, the local populace knows the social and environmental condition within which it must function. This is beyond the knowledge of outside experts. They will also have technical knowledge that may be vital to the effective functioning of any proposed interventions, built up from millennia of collective experience.

Water harvesting techniques are more likely to be adopted when all parties (both local and foreign) involved with the project are first brought to a round-table-discussion with the ultimate aim of finding the best technical option for the locality. Participants must include representatives of local administrative institutions, both formal and traditional. A key goal is to ensure that the systems proposed are appropriate to the locality and simple enough for the people to implement and maintain without outside assistance. Once this has been done, the chosen technique should be demonstrated, accompanied by some on-farm or on-station research. After obtaining some initial encouraging results, it is recommended to institutionalize the whole process of the water harvesting development in order to ensure its continuity. Incentives may be needed to encourage adoption of the proposed intervention; these may be in the form of tools or other necessary implements, motivational campaigns, good extension services, and training (Critchley & Siegert, 1991).

If investments (e.g. into storage basins, larger structures, canals) are necessary, an additional support by micro-finance might be needed (Nijhof & Shrestha, 2010).

9.6 WATER HARVESTING AND SUSTAINABILITY IN AGRICULTURE

For millennia, agriculture in dry areas of the world has been faced with the uncertainty and unpredictability of rainfall. In modern times three more problems must be faced by the farmers of water-scarce regions (FAO, 1997; UNEP, 2009):

1. The tremendous population growth with its unforeseeable impact on the resources.
2. The ever increasing competition for water resources between agriculture and urban areas.
3. The consequences of global climate change.

   These combine to make rainfed farming there very risky. Scientists and farmers are proposing a number of ways to deal with the problem of increased water demand and diminishing availability, including increasing water-use efficiencies, supplemental irrigation, and the use of drainage, waste, or even saline water.

The need for sustainability is stressed whenever development is discussed. If it is to be more than a mere rhetorical term, sustainability has to be properly defined. The Technical Advisory Committee (TAC) for the Consultative Group on International Agricultural Research (CGIAR) defined sustainable development as: "... successful management of resources for agriculture to satisfy changing human needs while maintaining or enhancing the quality of the environment and conserving natural resources" (Walsh, 1991). Walsh (1991) emphasized that "Agriculture [in the developing countries], therefore, faces a double challenge – not simply to increase food production, but to assure that the resource base is not degraded". How then can the goal of sustainable agriculture in dry areas be achieved? Under very harsh conditions, there is no easy answer on how resources could be successfully managed to:

- satisfy human needs, i.e. produce agricultural crops and livestock economically, while creating job opportunities, etc.;
- maintain or enhance the quality of the environment (including soil);
- conserve natural resources (including vegetation, especially trees and bushes which are often needed as sources of food or fuel).

Sustainability in agriculture is a goal that has rarely been reached. Nevertheless, it is imperative to aim for it in order to enhance environmental quality and the resource base on which agriculture depends. It is important to note that about two-thirds of the water consumed around the world goes to agriculture. Thus agriculture must play its part in conserving water resources.

There are numerous aspects of sustainability related to water harvesting systems (Figure 9.4); these are discussed in detail below.

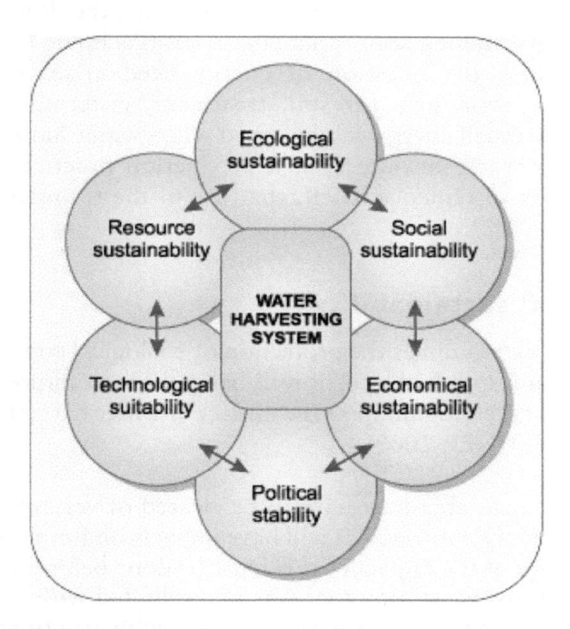

Figure 9.4 The various aspects of sustainability in regard to water harvesting (Prinz, 1994).

9.6.1 Resource sustainability

In the context of resources, 'sustain' means to maintain or prolong the productive use of resources and the integrity of the resource base. This has many implications for water harvesting systems.

- **Land use:** First, one must look at how much land is available for water harvesting and whether it is acceptable to leave an area uncropped as a runoff area (catchment). The optimum catchment: cropping area ratio must be determined.
- **Soils:** The soil characteristics need to be determined and compared. Maintaining a high level of soil fertility by fertilizing the run-on area also comes into question. Catchment's soils may need to be treated physically or chemically. And the long lasting consequences must be analyzed.
- **Water:** Climatic data for the area in question should be checked for reliability. Is the rainfall characteristic in favor of water harvesting? There may be other sources of water available, such as surface water and groundwater. To what extent they are available in terms of quantity, quality and price need to be ascertained. The conjunctive use of water resources in order to achieve a higher productivity security also comes into question. What water storage medium (soil, reservoir, and tank) is best suited? One must determine the benefits of building a reservoir, cistern or tank. Spillways for the evacuation of surface waters may be necessary. Does water harvesting go along with 'nutrient harvesting'? The need exists to measure the distance between the catchment and the run-on areas. Diversion structures are sometimes required. When downstream of the *Wadi*, no or little water runs when

water harvesting is practiced upstream, this is called the downstream effect and plays a role in determining appropriate measures (Pachpute *et al.*, 2009).

- **Energy:** Of course the question of energy need arises when constructing water harvesting structures for soil, treatment, cisterns, tanks etc. On the other hand, how much energy can be saved when water harvesting is employed instead of other water resources? If a production practice "takes a resource beyond its ability to replenish itself, that use of the resource would be unsustainable" (Faeth, 1993).

9.6.2 Ecological sustainability

'Ecological sustainability' implies the protection of ecological conditions necessary to support human life at a specified level of well-being through future generations (Lélé, 1991). Care must be taken to minimize the impact of water harvesting harm the local flora, fauna, and soil (UNEP, 2009).

- **Flora:** The catchment area may have to be cleared of vegetation, compacted, or chemically treated. Obviously, this will have impacts on flora growing in that area and an assessment of the impact of this must be done before starting work. Also, what happens to the vegetation that was naturally fed with runoff water before the project was started? A water harvesting project in arid or semiarid areas may result in the loss of genetic resources, and this must be evaluated.
- **Fauna:** The environmental changes imposed by a water harvesting scheme may affect wildlife and other fauna, for example by reducing their access to feed or water or interrupting their migratory routes. These impacts should be assessed and remedial measures taken if necessary (See also para 10.6).
- **Soil erosion/desertification:** If the catchment area is cleared of vegetation, more water and wind erosion may occur, perhaps even leading to desertification. Appropriate measures should be taken to combat these negative effects.

9.6.3 Social sustainability

Social sustainability is defined as "the ability to maintain desired social values, traditions, institutions, cultures, or other social characteristics" (Barbier, 1987). But social sustainability in the context of agricultural development should also include individual welfare, societal welfare, and equity. Introducing a new agricultural technique (such as water harvesting in many areas) requires a socially-appropriate implementation strategy, and assessment of the work-load involved, the risk aversion and management abilities of potential users, and the likely occurrence of new human (waterborne) diseases, among other things (Siegert, 1994).

Some of the many issues to be addressed include the following:

- **Tradition:** Some regions have used or use traditional techniques of water harvesting. How useful have these been, and are they still appropriate?
- **People's priorities:** The introduction of water harvesting into an area should meet local people's needs and priorities. Are they willing to take on the additional

labor involved? Their view of the risk of failure will affect their willingness to be involved in the project.

- **Participation:** Those benefiting from the project should also be involved in its development in terms of planning, surveying, construction, maintenance, monitoring, etc. Do they need assistance and, if so, what kind?
- **Gender and equity:** One group of people or one sex may benefit most or even solely from water harvesting. This issue needs to be addressed because of the long-term influence such an inequality could have on sustainability.
- **Land tenure:** Farmers may have individual rights to the land they cultivate or it may be communal or state land. These land rights must be taken into account in deciding where a water harvesting project is to be implemented.
- **Diseases:** Measures may need to be taken to prevent the spreading of water-borne diseases when runoff water is stored in tanks and ponds.

9.6.4 Other sustainability aspects

9.6.4.1 *Economic sustainability*

Higher yields or better growth of domestic animals and a higher income resulting from an intervention should pay for the additional investment/labor input for water harvesting. Ideally, the use of low-grade agricultural land could be intensified by water harvesting to give higher output and provide new job opportunities.

9.6.4.2 *Technological sustainability*

"Numerous water harvesting projects have failed because the technology used turned out to be unsuitable for the specific conditions of the site" (Siegert 1994). This might refer to a higher level of technology, connected with a higher input and management level, or to a technology that is not compatible with the local lifestyle.

9.6.4.3 *Political sustainability*

Sustainable development requires favorable, stable political and institutional conditions. Stable economic conditions, a well-developed infrastructure, government support services, e.g. a functioning extension service, and secure land rights are some of the most important issues in this respect (Tauer & Prinz, 1992; FAO, 1997).

Resources sustainability, ecological sustainability, and social sustainability' are of primary importance, the other three of lesser importance, but all six have to be duly considered as they are all interrelated with each other (The goals of the various forms of sustainability are not always compatible with each other). On the other hand, the various aspects of sustainability have a synergetic effect: The planting of shelterbelts supported with runoff water from micro-catchments (Boers, 1994) improves the ecological resource and economic sustainability of the area.

Despite considerable efforts in recent years to promote and disseminate water harvesting in drought-prone areas, the overall success is much less than expected. It is seldom that the technology itself is fundamentally wrong, but rather

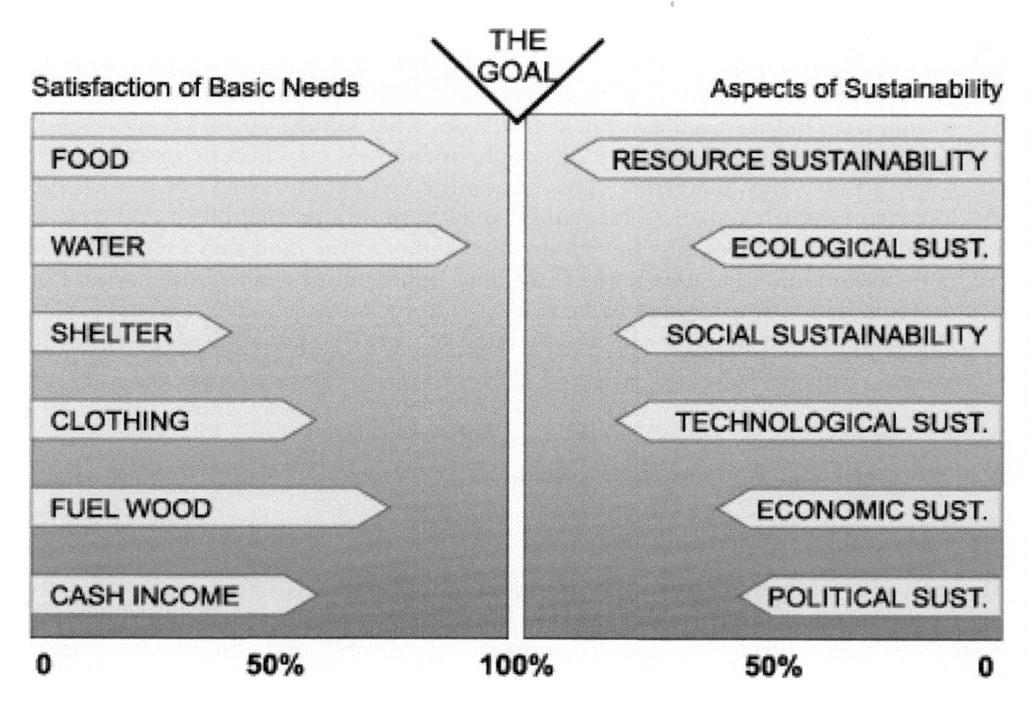

Figure 9.5 Hypothetical example for a sustainability achievement assessment (Prinz, 1994).

the selected method does not correspond with the specific situation of the site and often it turns out that the chosen implementation approach is unsuitable for the conditions of the area (Siegert, 1994). The meeting of human needs and all aspects of sustainability have to be viewed simultaneously and the degree of achievement clearly stated (Figure 9.5).

Sustainable achievement assessments (Figure 9.5), comparable to a certain degree with environmental impact assessments, should be carried out very sensitively and thoroughly to avoid the many mistakes made in the past.

CHAPTER 10

Water quality and environmental considerations

10.1 INTRODUCTION

Water is never found in pure state in nature. Essentially, all water contains substances derived from the natural environment and human activities. These constituents determine water quality. Water quality is a prime factor in determining the suitability of water supplies to satisfy the requirements of different uses. Storing water in tanks, reservoirs, etc. poses quality and hygienic problems, especially in warmer climates. Thus, water quality considerations differ between micro-catchment and macro-catchment systems, and between systems with and without interim storage.

The implementation of water harvesting systems has numerous impacts on the environment, e.g. on aquatic life, and also on the spread of water-related diseases.

This chapter briefly covers all these aspects, with emphasis on agricultural uses of harvested water.

10.2 WATER HARVESTED FOR HUMAN CONSUMPTION

Water for domestic uses and drinking must meet certain qualitative standards. WHO (2004) provide guidelines for the provision of safe drinking water, including quality standards and information on the roles and responsibilities of various stakeholders involved in providing drinking water. However, the practical application of these requirements varies from place to place depending on the living standard of the community and type of water source.

The main water quality indicators of drinking water are characterized by their physical, chemical, and biological parameters (RAIN, 2008). The list of the main indicators/parameters includes:

- *Alkalinity*
- *Color of water*
- *pH*
- Taste and odor
- Dissolved metals and salts (*sodium, chloride, potassium, calcium, manganese, magnesium*)
- Microorganisms such as *fecal coliform bacteria* (*Escherichia coli*), *Cryptosporidium*, and *Giardia lamblia*

- Dissolved metals and metalloids (*lead, mercury, arsenic*, etc.)
- Dissolved organics: *colored dissolved organic matter, dissolved organic carbon, etc.*
- *Heavy metals*

Therefore, the basic requirements for safe drinking water are:

- free from disease-causing organisms;
- free from compounds that have an adverse effect on human health;
- fairly clear (low turbidity and little color); and
- without offensive taste or smell.

There are several simple methods for treating water before drinking to improve its biological properties:

- Boiling will kill any harmful bacteria that may be present.
- Adding chlorine in the right quantity (35 ml of sodium hypochlorite per 1000 liters of water) will disinfect the water.
- Slow sand filtration will remove any harmful organisms when carried out properly (RAIN, 2008).
- Using the flocculation power of Moringa tree's seeds to purify water (About 100 mg per liter water).
- Use of sunlight (ultraviolet compounds) to disinfect water filled in plastic tubes.

Further information on water disinfection and fresh-keeping of stored rainwater is found in Zhu *et al.* (1999).

10.3 WATER HARVESTED FOR ANIMAL CONSUMPTION

Livestock generally require water similar to that which would be suitable for human consumption, although many animals tolerate water of somewhat poorer quality. Farm water supplies should be protected against contamination with microorganisms, chemicals, and other pollutants. Substances that originate on livestock farms and often contaminate water supplies include nitrates, bacteria, and organic materials, including feces, and suspended solids. A high level of suspended solids and an objectionable taste, odor or color in water can cause animals to drink less than they should.

Measures must be taken to prevent human and animal wastes contaminating the water resource. Most of the bacterial contamination of water occurs at or after water has been collected. Similarly any water spilled at the collection point which drains back to the water body is almost certain to contain bacterial contaminants. Household wastewater is often returned to water courses without any treatment. When this wastewater becomes a significant part of the total flow, the bacterial quality of the resource quickly declines, and it can become a serious health hazard for both animals and humans. Great care (and awareness) is needed to prevent these and many other simple causes of drinking water contamination (Figure 10.1). The trough to water livestock should be located at the opposite side of the cistern than the inflow opening. Otherwise the droppings of the animals are carried by the runoff straight into the cistern.

Figure 10.1. Cistern in NE Libya. The water from the cistern is pumped into the trough (foreground) to supply goats, sheep, and other animals. To prevent contamination of the cistern water, the trough should be located several meters apart and the inflow opening at the opposite side. Photo courtesy D. Prinz/Karlsruhe University, Germany. (*See color plate, page 258*).

Surface water supplies to which livestock have ready access are always potential candidates for contamination. The presence of coliform bacteria in a well is an indication that surface water is finding its way into the well. Water can serve as a home for many different disease organisms and toxins. Stagnant water contaminated with manure or other nutrients may develop blue-green algae, which can poison livestock. Farm pond water needs to be observed for the presence of algae and other harmful organisms during hot, dry weather. The organism can survive for extended periods in surface waters. One should take care to avoid forcing livestock to drink from water sources that may be contaminated with urine.

When water is suspected of causing health problems in livestock, veterinary assistance should be sought to determine the actual disease. Laboratory diagnostic examination of animals as well as the water supply may be necessary to determine the cause. Temporarily changing to a known safe water supply is a useful test to determine whether the health problems can be solved. Water is too often blamed for production or disease problems. Thus, the importance of an accurate diagnosis must be emphasized.

The US Department of Agriculture (1992) recommends that livestock water contain less than 5,000 coliform organisms per 100 milliliters; fecal coliform should be near zero (Table 10.1).

Nitrates are soluble and move with percolating or runoff water. Therefore, ponds with runoff from heavily fertilized or manured fields may contain high levels of nitrates. Nitrate nitrogen is not especially toxic. Ruminants can convert some nitrate to usable products. Nevertheless, more than 300 parts per million (ppm) nitrate nitrogen (NO_3-N) may cause nitrate poisoning in cattle, sheep, or horses, and

Table 10.1. Desired levels of pollutants in livestock water supplies, and levels at which problems are likely to occur.

Substance	*Desired range*	*Problem range*
Total bacteria per 100 milliliters	<200	>1 000 000
Fecal coliforms per 100 milliliters	<1	>1 for young animals >10 for older animals
Fecal streptococci per 100 milliliters	<1	>3 for young animals >30 for older animals
pH	6.8 to 7.5	<5.5 or >8.5
Dissolved solids, milligrams per liter	<500	>3000
Total alkalinity, milligrams per liter	<400	>5000
Sulfate, milligrams per liter	<250	>2000
Phosphate, milligrams per liter	<1	Not established
Turbidity, Jackson units	<30	Not established

Source: U.S. Department of Agriculture (1992).

its use for these animals is not recommended. Because this level of nitrate contributes to the salts content in a significant amount, use of this water for swine or poultry should be avoided.

If there is any sign of a negative effect of drinking water on animal health, a water test is recommended. Sample bottles should be obtained from the testing laboratory or local health department, because containers may be especially prepared for a specific contaminant. Sampling and handling procedures depend on the water-quality concern and should be followed carefully.

10.4 WATER HARVESTED FOR CROP PRODUCTION

The quality of water used for irrigation is an important factor in productivity and agricultural sustainability (Ayers & Westcot, 1994). In irrigation water evaluation, emphasis is placed on the chemical and physical characteristics of the water and only rarely is any other factor considered important. In most locations where water harvesting for agriculture is practiced, the physical quality of water is much more important than the chemical quality. Of importance are the quantities of solids transported and where they are deposited, and the nutrients, and in rare cases also pollutants, carried by the particles (Figure 10.2).

If the sediments are rich in clay, water infiltration may be hampered. Relatively high sodium or low calcium content of soil or water also reduces the rate at which irrigation water enters soil. If too little water infiltrates the soil (and more water evaporates), the crop may not survive from one rainfall event to the next.

Some of the chemicals used to treat catchments, or the degradation products of these chemicals, may affect plant growth, but reports are lacking. If the runoff water runs over salt-bearing rocks it can be contaminated with salt, reducing crop growth.

Figure 10.2 A flooded *wadi* in southern Morocco. Each flooding is associated with sediment transport increasing the turbidity of the water. Photo courtesy D. Prinz/Karlsruhe University, Germany. (*See color plate, page 259*).

10.5 WATER QUALITY CONSIDERATIONS RELATED TO WATER HARVESTING METHODS

Each type of water harvesting system is associated with particular types of water-quality problems (Table 10.2).

10.5.1 Rooftop and courtyard systems

These systems are normally equipped with some kind of storage (tanks, jars, barrels, etc.) (Table 10.2, Figure 10.3). Water collected from a roof, courtyard, etc. may contain biological, chemical, and physical impurities (Gould & Nissen-Petersen, 1999; Heijnen & Pathak, 2007; RAIN, 2008).

Generally the quality of rainwater, if collected in a clean vessel, falls within the WHO guidelines and rarely present problems.

There are several key considerations when looking at the quality and health aspects of harvesting rainwater from rooftops and courtyards. First, there is the issue of bacteriological water quality. Rainwater can become contaminated by feces (e.g. bird droppings) entering the tank from the catchment area. This can be avoided by regular and thorough cleaning of the catchment surface. Rainwater tanks should be designed to prevent contamination by leaves, dust, insects, vermin, and industrial or

Table 10.2 Major water-quality considerations according to type of water harvesting system.

	Kind of catchment	*Kind of use*	*Type of storage*	*Major water-quality considerations*
Rooftop and courtyard systems	Treated surfaces (e.g. sealed, paved, compacted, smoothened surfaces)	Drinking water for humans and domestic animals, irrigation of vegetables, ornamental plants	Cisterns, ponds, jars, tanks	Contamination by bird droppings Contamination by corroding/degrading material Dust and dirt accumulation Regular cleaning necessary
On-farm system	Usually treated	Trees, bushes, annual crops	Soil profile, ponds	Sedimentation of run-on area/planting basin Nutrient inflow from catchment area Crop damage due to agricultural chemicals
Long-slope water harvesting	Treated or untreated	Trees, bushes, annual crops, drinking water for domestic animals and humans	Soil profile, cisterns, ponds, reservoirs	Sedimentation of distribution basins and canals Loss of storage capacity Nutrient inflow from catchment Salinity
Floodwater harvesting	Untreated	Trees, bushes, annual crops	Soil profile, ponds, reservoirs	Sedimentation of distribution basins and canals Loss of storage capacity Nutrient inflow from catchment

Figure 10.3 Rainwater tank in Amhara region, Ethiopia. The water collected from the rooftop is used mainly for domestic purposes, sometimes for gardening. Photo courtesy E.G. Zerihun/ Water Resources Development Program, Organization for Rehabilitation and Development in Amhara (ORDA), Bahir Dar, Ethiopia. (*See color plate, page 259*).

agricultural pollutants. Tanks should be sited away from trees, with well-fitted lids, and kept in good condition. Incoming water should be filtered, screened, or allowed to settle to remove foreign matter before entering the storage facility. Water that is relatively clean on entry to the tank will usually improve in quality if allowed to settle for some time inside the tank. Bacteria entering the tank will die rapidly if the water is relatively clean. Keeping a tank dark and in a shady spot will prevent growth of algae and also keep the water cool.

Second, there is the problem of chemical pollution. This can have its origin in a number of sources:

- Corroding/degrading material from the catchment area: Metal roofs corrode; asphalt and certain plastics are degraded by sunlight and heat, releasing water-soluble substances. A thatched roof is normally unsuitable for water harvesting if the water is to be consumed by humans or domestic animals.
- Air pollution from industrial emissions, deposited wet or dry on the rainwater catchment area. Rain may be highly acidic (pH of about 4.3), primarily due to dissolved sulfates and nitrates. Trace amounts of heavy metals have also been identified in rain. Higher concentrations are usually found around industrial areas.

- Automobile emissions: Exhaust emissions and oil can contaminate water collected from road surfaces or surfaces near streets with high traffic density.

The physical quality of water will also be affected by dust and dirt; organisms and water-soluble impurities from windblown dust can affect the biological quality. Fine sediments, particularly those of colloidal size, are often a source of contamination. The sediments themselves may be harmless but many forms of contaminants, both bacterial and dissolved salts, often adhere to the microscopic particles. Allowing water to stand for a few days and then consuming only the clear water from the top can avoid many of the dangers.

A number of actions can be taken to ensure the quality of water harvested from rooftops and courtyards:

- The water can be filtered through sand or membranes to remove all particulate contamination but this is expensive and requires some specialized equipment. Also, filtering is totally ineffective in removing dissolved contaminants such as various salts (NaCl, $CaCO_3$) that have not adhered to the surfaces of particulates.
- After a longer dry period, organic matter, e.g. bird droppings, dust, and debris will have accumulated on the catchment area. To avoid this being washed into the tank, the 'first flush' water should be diverted so that it does not enter the tank. There are a number of ways of doing this. The simplest way is a manually operated arrangement whereby the inlet pipe is moved away from the tank inlet and then replaced once the initial first flush has been diverted. This method has obvious drawbacks in that there has to be a person present who will remember to move the pipe. Other systems use tipping gutters to achieve the same purpose. However, even better would to clean the gutters and roofs or harvesting area before the onset of the rainy season.
- Tanks should be sealed to prevent insects from entering. Mosquito proof screens should be fitted to all openings. Some practitioners recommend putting 1 to 2 teaspoons of household kerosene in a tank of water, which provides a film that prevents mosquitoes settling on the water and suffocates any larvae present.
- The tank should be emptied at the end of the dry season and any silt collected in the bottom should be removed. The interior of the tank should be cleaned thoroughly. Gaps between roofs and walls should be closed with stone and mortar to keep out windblown materials, lizards, birds, and insects.
- The ropes and bucket used to raise water from underground tanks or cisterns should be kept clean, and those handling them should wash their hands to avoid introducing pollutants (Zhu *et al.*, 1999).

These suggestions refer not only to storage tanks but also to cisterns (Figure 10.4).

The area surrounding a rainwater harvesting system should be kept in good sanitary condition and fenced off to prevent animals fouling the area or children playing around the tank. Any pools of water gathering around the tank or cistern should be drained and the depressions filled.

Figure 10.4 A constructed catchment with cistern to cover the drinking and domestic water demand of a small Tunisian village. Photo courtesy D. Prinz/Karlsruhe University, Germany. (*See color plate, page 260*).

10.5.2 Runoff water from on-farm micro-catchment systems

Runoff water from the catchment is used directly (i.e. without interim storage) to irrigate crops. Here, nutrients in the runoff water will be beneficial, but sediments may reduce the physical quality of the water. If the catchment areas were chemically treated, e.g. with herbicides, salt, silicone, tar, oil, etc., these can pose a threat to the crop plants.

10.5.3 Long-slope water harvesting

Runoff water from long slopes is used to provide additional water to trees, bushes, or annual crops on the cropped area. In most cases the water is conserved directly in the soil profile, although it is sometimes stored in cisterns, ponds, or reservoirs. In this case the collected water may be used to water livestock or for domestic purposes, too. The catchments are either left in a natural state or cleared of vegetation and stones; in either case there is a high risk of soil erosion and hence of sediment transport to stream channels and into the storage bodies.

For small storage facilities, it may be possible to construct a sediment trap upstream of the reservoir (Oweis *et al.,* 1999). Within a few days most suspended sediments settle at the bottom of the trap and the clear surface water can be directed to the main storage facility (Figure 10.5). Some of the runoff water will be lost from

Figure 10.5 In this example from Ethiopia, runoff water has to pass through two settling basins before flowing into the sealed pond. Photo courtesy G. Zerihun/ORDA, Bahir Dar, Ethiopia. (*See color plate, page 260*).

the sediment trap but this may be a small price to pay for increasing the life of the storage facility.

The best strategy for dealing with sediments is to prevent water and wind erosion.

Animals and people should be kept away from water stored in ponds or reservoirs using secure fencing or hedges.

Water for animals should be supplied in a drinking trough. Allowing animals to drink directly from a surface pond results in contamination of the water.

10.5.4 Floodwater harvesting

Floodwater harvesting is commonly used to supply water to trees, bushes, and annual crops. In a number of cases the water is stored in ponds and reservoirs. The water moves soil particles, which may carry nutrients as well as chemical pollutants (Figures 10.6 and 10.7). Sedimentation of distribution basins, canals, and storage bodies are consequence common problem because of the large amount of sediment carried by floodwater. The ephemeral river bed or *wadi* will also be affected.

Deposition of sediment changes the character of the *wadi*. The sediments build up on the *wadi* bed, reducing the channel cross-section, and changing the hydraulic gradient. This can have effects both upstream and downstream, with the *wadi* increasingly likely to overflow its banks during high flows. Banks may be eroded and the *wadi* may even change its course (Figure 10.8). During floods large amounts of

Figure 10.6 Flow of water after a rainfall event to an ephemeral river bed in southern Tunisia. The water moves soil particles, which may carry nutrients as well as chemical pollutants. Photo courtesy B. Chahbani/IRA, Medenine, Tunisia. (*See color plate, page 261*).

Figure 10.7. Runoff water flowing through a culvert after a rainfall event in the Syrian steppe. The water is rich in sediments, which settle further downstream. Photo courtesy T. Oweis/ICARDA. (*See color plate, page 261*).

Figure 10.8 The sediments carried by the flowing water will settle and change the morphology of the ephemeral river course. Photo courtesy D. Prinz/Karlsruhe University, Germany. (*See color plate, page 262*).

sediment may be deposited on the floodplain, which may be beneficial in the long term but may be undesirable in the short term.

The more obvious problem produced by sediments is the loss of storage capacity of reservoirs. It is not uncommon for storages to be completely filled with sediment within a few years of construction.

One efficient means to trap sediments is by constructing rock dikes across valleys to capture surface water. The sediments will settle and create a flat surface for crop growing, particularly trees with deep roots to reach the water stored in the trapped sediments, as in the case of the *jessour* in North Africa (See chapter 3).

Stored water needs to be protected not only from evaporation and seepage loss but also from contamination. Contamination occurs mainly from human or other animal contact. Similarly, stored water needs to be protected from disease vectors such as mosquitoes, flies, and mollusks.

10.6 IMPACTS ON DOWNSTREAM ECOSYSTEMS AND BIODIVERSITY

Historically, aquatic flora and fauna have developed in concert with existing water regimes. Implementing a water harvesting system alters the flow regime, which will cause some species to become stressed and die off. Other species that prefer the new flow regime will then colonize the *wadi* bed and flood-plain. These changes may be seen as desirable or undesirable. In most locations, not only are the aquatic species adapted to the flow regime, but other species, including humans, have adapted their lifestyles to be at least partially dependent on the existing environmental conditions. In recent years it has been realized that many of the changes produced by significant modification of flow regimes are undesirable. Policies have been developing to reduce the changes in the flow regimes such that the general environment downstream will not be drastically modified.

Downstream degradation of the aquatic flora and fauna produced by cutting off all flows will eventually be drastic and irreversible.

In most situations, the aquatic environments are not sufficiently understood for clear guidelines to be developed. Post development flows are being set at about 25% of the pre-development situation, with similar variability, but this is an arbitrary figure which has no scientific basis. Presumably it will be enough to support some species, but others will disappear. Environmental flows, the flows needed to sustain the naturally occurring species (flora and fauna), are very difficult to define and are currently the subject of vigorous scientific and political debates (Pereira, 2002). For many streams, maintenance of the flows necessary to support the natural systems means virtually no development can occur. In other regimes, particularly in semi arid and arid regions, very little is known about possible effects of developing the very limited water resources. It is quite possible that every region would be very different. These effects may not be immediately apparent, but in the long run they will be noticed, but by then it will probably be impossible to reverse the effects even if there was a desire to do so.

However, water harvesting projects can also contribute to higher levels of biodiversity, e.g., by having small areas (each of about 100 m^2) fenced at various locations to allow the natural flora to develop and set seeds, undisturbed by grazing animals and well supplied by harvested water.

10.7 WATER-BORNE DISEASES

Any form of water-resource development, such as a water harvesting intervention, causes changes in natural conditions. Many of these changes offer opportunities for multiplication of disease vectors, with devastating effects.

In regions where malaria, dengue fever, or similar insect-transmitted diseases are endemic, storage of water on the surface needs to be accompanied by precautionary measures to prevent the water becoming a breeding site for these disease vectors. Where schistosomiasis is prevalent, measures must be taken to control the snail that is the intermediate host of the parasite.

It is important that planners, decision-makers and financiers take health issues into consideration when planning of any water-resource development project. This will often require changes to the scheme and may raise the cost of the project. But with innovative ideas the changes and extra costs need not be large. Examples are to be found in many irrigation developments in regions where *wadi or river blindness* and *schistosomiasis* are endemic.

References

ACPC (African Climate Policy Center of United Nations Economic Commission for Africa—UNECA) (2011). *Agricultural water management in the context of climate change in Africa*. Working Paper 9, UNECA-ACPC, Addis Ababa, Ethiopia. (http://www.uneca.org/acpc/publications)

Agarwal, A. & Narain, S. (1997). *Dying Wisdom. Rise, fall and potential of India's traditional water harvesting systems*. Centre for Science and Environment (CSE), New Delhi, India.

Ahmad, S. (2004). Indigenous water-harvesting systems in Pakistan. In: Oweis, T., Hachum, A. & Bruggeman, A. (eds.) *Indigenous water harvesting systems in West Asia and North Africa*. International Center for Agricultural Research in the Dry Areas (ICARDA), Aleppo, Syria. pp. 151–175.

Alghariani, S.A. (1994). Contour ridge terracing water harvesting systems in northwestern Libya: the *Amamra* project as a case study. In: *Water harvesting for improved agricultural production: Proceedings of the FAO Expert Consultation, Cairo, Egypt, 21–25 November 1993*. Food and Agriculture Organization of the United Nations, Rome, Italy. FAO Water Reports 3. pp. 35–57.

Ali, A., Bulad, A., Kozah, A., Oweis, T. & Bruggeman, A. (2006). Fighting desertification in Jordan and Lebanon. *International Center for Agricultural Research in the Dry Areas, ICARDA Caravan*. [Online] 23. Available from: http://www.icarda.org/Publications/Caravan/Caravan23/Focus_2.htm [accessed 23th December 2006].

Ali, A., Oweis, T., Salkini, A.B. & El-Naggar, S. (2009). *Rainwater cisterns: traditional technologies for dry areas*. International Center for Agricultural Research in the Dry Areas (ICARDA), Aleppo, Syria.

Allen, R.G., Pereira, L.S., Raes, D. & Smith, M. (1998). *Crop evapotranspiration. Guidelines for computing crop water requirements*. Food and Agriculture Organization of the United Nations, Rome, Italy. FAO Irrigation and Drainage Paper 56.

Al-Salaymeh, A., Al-Khatib, I.A. & Arafat, H.A. (2011). Towards sustainable water quality: management of rainwater harvesting cisterns in southern Palestine. *Water Resources Management*, 25 (6), 1721–1736.

Antinori, P. & Vallerani, V. (1994). Experiments in water harvesting technology with the dolphin and train ploughs. In: *Water harvesting for improved agricultural production*. Expert Consultation, Cairo, Egypt 21–25 November 1993. Food and Agriculture Organization of the United Nations, Rome, Italy. pp. 113–132.

Ayers, R.S. & Westcot, D.W. (1994). *Water quality for agriculture*. Food and Agriculture Organization of the United Nations, Rome, Italy. FAO Irrigation and Drainage Paper 29 Rev.

Bamatraf, A. (1994). Water harvesting and conservation systems in Yemen. In: *Water harvesting for improved agricultural production*. Expert Consultation, Cairo, Egypt, 21–25 November 1993. Food and Agriculture Organization of the United Nations, Rome, Italy. pp. 169–189.

Barbier, E.B. (1987). The concept of sustainable economic development. *Environmental Conservation,* 14 (2), 101–110.

Barrow, C. (1987). *Water resources and agricultural development in the tropics.* Longman, UK.

Barrow, C. (1999). *Alternative Irrigation. The promise of runoff agriculture.* Earthscan Publ.

Bates, B.C., Kundzewicz, Z.W., Wu, S. & Palutikof, J.P. (eds.) (2008). *Climate change and water.* Intergovernmental Panel on Climate Change Secretariat, Geneva, Switzerland. IPCC Technical Paper VI.

Bazza, M. & Tayaa, M. (1994). Operation and management of water harvesting techniques. In: *Water harvesting for improved agricultural production: Proceedings of the FAO Expert Consultation, Cairo, Egypt, 21–25 November 1993.* Food and Agriculture Organization of the United Nations, Rome, Italy. pp. 271–286.

Ben Mechlia, N. & Ouessar, M. (2004). Water-harvesting systems in Tunisia. In: Oweis, T., Hachum, A. & Bruggeman, A. (eds.) *Indigenous water harvesting systems in West Asia and North Africa.* International Center for Agricultural Research in the Dry Areas, Aleppo, Syria. pp. 21–41.

Ben Mechlia, N., Oweis, T., Masmoudi, M., Khatteli, H., Ouessar, M., Sghaier, N., Anane, M. & Sghaier, M. (2009). *Assessment of supplemental irrigation and water harvesting potential—Methodologies and case studies from Tunisia.* International Center for Agricultural Research in the Dry Areas, Aleppo, Syria.

Ben-Asher, J. & Berliner, P.R. (1994). Runoff irrigation. In: K.K., Tanji & B., Yaron (eds.). *Advances Series in Agricultural Sciences* 22, pp. 126–154.

Ben-Asher, J., Prinz, D., Laron, J. & Abravia, I. (1995). Greenhouse roof top water harvesting: A case study under Mediterranean conditions. In: *Water resources management in the Mediterranean under drought or water shortage conditions: Proceedings of the European Conference on Water Use, Nicosia, Cyprus, 14–18 March 1995.* Balkema, Rotterdam. pp. 145–152.

Berliner, P.R. (2005). Runoff agriculture. In: U. Aswathanarayana (ed.). *Advances in Water Science Methodologies.* Leiden, The Netherlands: Balkema Publ.

Beven, K.J. (2000). *Rainfall-runoff modeling.* New York, The Primer, John Wiley & Sons, Ltd.

Boers, Th.M. (1994). *Rainwater harvesting in arid and semi-arid zones.* International Institute for Land Reclamation and Improvement, Wageningen, The Netherlands. Publication 55.

Borst, L. & Haas, S.A. de, (2006). *Hydrology of sand storage dams. A case study in the Kiindu catchment, Kitui District, Kenya.* Amsterdam, NL: Free University Amsterdam/ACACIA Institute.

Bruins, H.J., Evenari, M. & Nessler, U. (1986). Rainwater-harvesting agriculture for food production in arid zones: the challenge of the African famine. *Applied Geography* 6, 13–32.

Carmi, G. & Berliner, P. (2008). The effect of soil crusting on the runoff generation of small plots in an arid environment. *CATENA* 74:37–42.

Carruthers, I.D. (1983). *Aid for the development of irrigation.* Report on the development assistance committee workshop on "Irrigation Assistance" held in Paris on 29–30 September 1982. Paris: OECD (Organization for Economic Cooperation and Development).

CGWB, (2011). *Rainwater harvesting and artificial recharge.* Selected case studies. Central Ground Water Board, Ministry of Water Resources, New Delhi, India.

Chow, V.T., Maidment, D.R. & Mays, L.W., (1988). *Applied Hydrology.* McGraw-Hill Inc.

Cluff, C.B. (1975). Engineering aspects of water harvesting research at the University of Arizona. In: G.W. Frasier, (ed.) *Proceedings of the Water Harvesting Symposium, Phoenix, Arizona, 26–28 March 1974.* US Department of Agriculture, Agricultural Research Service, Berkley, California, USA. pp. 27–39.

Critchley, W. & Siegert, K. (1991). *Water harvesting: A manual for the design and construction of water harvesting schemes for plant production.* Food and Agriculture Organization of the United Nations, Rome, Italy. Available from: http://www.fao.org/docrep/U3160E/u3160e00.htm (Date of access 20th Feb. 2012).

Critchley, W., Reij, D. & Seznec, A. (1992). *Water harvesting for plant production. Vol. II: Case studies and conclusions for sub-Saharan Africa.* World Bank, Washington, D.C., USA. World Bank Technical Paper 157.

Cullis, A. (1987). Turkana water harvesting project (KEN 197). In: D. Reij, A. Cullis,. & J. Aklilu. *Soil and water conservation in sub-Saharan Africa*, Annex 3.

Cullis, A. & Pacey, A. (1992). *A development dialogue: rainwater harvesting in Turkana.* Intermediate Technology Publications, London, UK.

Davy, E.G., Mattei, F. & Solomon, S.I. (1976). *An evaluation of climate and water resources for development of agriculture in the Sudano-Sahelian zone of West Africa.* Special Environmental Report No. 9, World Meteorological Organisation, WMO-No. 459, Geneva, CH.

Dedrick, A.R. (1975). Storage systems for harvested water. In: G.W. Frasier (ed.) *Water Harvesting Symposium, March 26–28, 1974, Phoenix, Arizona.* US Department of Agriculture, Agricultural Research Service, Berkeley, California, USA. pp. 175–191.

De Pauw, E., Oweis, T. & Youssef, J. (2008). *Integrating expert knowledge in GIS to locate biophysical potential for water harvesting: methodology and a case study for Syria.* International Center for Agricultural Research in the Dry Areas, Aleppo, Syria.

Doolittle, W.E. (1984). Agricultural change as an incremental process. *Annals of the Association of American Geographers* 74, 124–137.

Doorenbos, J. (1976). *Agro-meteorological field stations.* Food and Agriculture Organization of the United Nations, Rome. FAO Irrigation and Drainage Paper No. 27.

Drolett, J., Shatanawi, M., Taimeh, A., Oweis, T. & El Jabi, N. (1997). *Development of Optimal Strategies for Water Harvesting in Arid and Semi Arid Lands.* Annual Reports of the IDRC-Jordan-Concodia-ICARDA-Moncton project.

Droppelmann, K. & Berliner, P.R. (2003). Runoff agroforestry—a technique to secure the livelihood of pastoralists in the Middle East. *Journal of Arid Environments.* 54: 571–577.

Dutt, G.R. (1981). Establishment of NaCl-treated catchments. In: G.R. Dutt, C.F. Hutchinson, and M. Anaya Garduno, (eds.). *Rainfall collection for agriculture in arid and semi-arid regions: Proceedings of a workshop, University of Arizona, Tucson, Arizona 10–12 September 1980.* Commonwealth Agricultural Bureaux, Farnham House, Farnham Royal, Slough, UK. pp. 17–21.

Eger, H. (1986). *Runoff agriculture: A case study about the Yemeni highlands.* Doctoral thesis, University of Tübingen, Germany.

El-Amami, S. (1983). *Les amenagements hydrauliques traditionels en Tunisie.* Centre de Recherche du Genie Rural, Tunis, Tunisia.

Erie, L.J., French, O.F., Bucks, D.A. & Harris, K. (1982). *Consumptive use of water by major crops in the southwestern United States.* US Department of Agriculture, Agricultural Research Service, Washington, D.C., USA. Conservation Research Report 29.

Evenari, M., Shanan, L. & Tadmor, N. (1982). *The Negev: the challenge of a desert.* Harvard University Press, Cambridge.

Faeth, P. (1993). Evaluating agricultural policy and the sustainability of production systems: an economic framework. *Journal of Soil and Water Conservation* 48, 94–99.

Falkenmark, M., Fox, P., Persson, G. & Rockström, J. (2001). *Water harvesting for upgrading of rainfed agriculture: problem analysis and research needs.* Stockholm International Water Institute (SIWI), Stockholm, Sweden, report number: 11.

FAO (Food and Agriculture Organization of the United Nations) (1977). *Soil conservation and management in developing countries.* Soil Bulletin 33. FAO, Rome, Italy.

FAO (Food and Agriculture Organization of the United Nations) (1997). *Guidelines for the integration of sustainable agriculture and rural development into agricultural policies.* Agricultural Policy and Economic Development 4. FAO, Rome, Italy.

FAO (Food and Agriculture Organization of the United Nations) (2001). *Water harvesting in western and central Africa.* FAO Regional Office for Africa, Accra, Ghana.

FAO (Food and Agriculture Organization of the United Nations) (2010). *Guidelines on spate irrigation.* [Online] FAO, Rome, Italy. Irrigation and Drainage Paper 65. Available from: *www.fao.org/docrep/012/i1680e/i1680e.pdf [Accessed 7th December 2010].*

Fardous, A.A., Taime, A. & Jitan, M. (2004). Indigenous water-harvesting systems in Jordan. In: Oweis, T., Hachum, A. & Bruggeman, A. (eds.). *Indigenous water harvesting systems in West Asia and North Africa.* International Center for Agricultural Research in the Dry Areas, Aleppo, Syria. pp. 42–60.

Fooladmand, H.R. & Sepaskhah, A.R. (2004). Economic analysis of the production of four grape cultivars using microcatchment water harvesting systems in Iran. *Journal of Arid Environments*, 58 (4), 525–533.

Frasier, G.W. (1980). Harvesting water for agriculture, wildlife and domestic uses. *Journal of Soil and Water Conservation* 35 (3), 125–128.

Gammoh, I.A. & Oweis, T.Y. (2011). Contour Laser Guiding for the Mechanized 'Vallerani' Micro-catchment Water Harvesting Systems. *Journal of Environmental Science and Engineering*, 5 (9).

Gould, J. & Nissen-Petersen, E. (1999). *Rainwater catchment systems for domestic supply: Design, construction and implementation.* ITDG Publications, London, UK.

GTZ/DGF (Deutsche Gesellschaft für Technische Zusammenarbeit/Direction Générale des Forêts). (1993). *Les tabias.* GTZ and Ministry of Agriculture, Tunis, Tunisia.

Hachum, A.Y. & Alfaro, J.F. (1980). *A physically-based model of water infiltration in soils.* Utah Agricultural Experiment Station, Logan, Utah, USA. Utah Agricultural Experiment Station Bulletin 505.

Hatibu, N., Mutabazi, K., Senkondo, E.M. & Msangi, A.S.K. (2006). Economics of rainwater harvesting for crop enterprises in semi-arid areas of East Africa. *Agricultural Water Management* 80, 74–86.

Heijnen, H. & Pathak, N. (2007). *Rain water harvesting quality, health and hygiene aspects: Proceedings of the Thirteenth International Conference on Rain Water Catchment Systems, Sydney, Australia August 2007.* The International Rainwater Catchment Systems Association.

Hogg, R. (1986). *Water harvesting in semi-arid Kenya: Opportunities and constraints.* Consultants' report for Ministry of Agriculture and Livestock Development, Nairobi, Kenya.

Hoogmoed, M. (2007). *Analyses of impacts of a sand storage dam on groundwater flow and storage.* Amsterdam, NL: Free University Amsterdam.

Hudson, N.W. (1987). *Soil and water conservation in semi-arid areas.* Food and Agriculture Organization of the United Nations, Rome, Italy. FAO Soils Bulletin 57.

Hurni, H. (1986). *Soil Conservation in Ethiopia.* Ministry of Agriculture, Addis Ababa, Ethiopia.

Hurni, H. (1989). Applied soil conservation research in Ethiopia. In: D.B. Thomas, E.K.Biamah, A.M. Kilewe, L. Lundgreen & B.O. Mochose (eds.) *Soil and water conservation in Kenya: Proceedings of the Third National Workshop.* Kabete, Nairobi, Kenya, 16–19 September 1986. pp. 5–21.

ICARDA (International Center for Agricultural Research in the Dry Areas) (2006). *Annual Report 2005.* International Center for Agricultural Research in the Dry Areas, Aleppo, Syria.

IPCC (Intergovernmental Panel on Climate Change) (2007). *Fourth assessment report: Climate change 2007*. Working Group II Report on 'Impacts, adaptation and vulnerability', IPCC, Geneva, Switzerland.

Israelsen, O.W., Stringham, G.E. & Hansen, V.E. (1980). *Irrigation principles and practices (4th ed.)*. John Wiley and Sons, Hoboken, New Jersey, USA.

Jensen, M.E. (ed.) (1980). *Design and operation of farm irrigation systems, ASAE Monograph 3*. American Society of Agricultural Engineers. St. Joseph, Michigan. USA.

Joshi, P.K., Jha, A.K., Wani, S.P. & Sreedevi, T.K. (2009). Scaling-out community watershed management for multiple benefits in rainfed areas. In: S.P. Wani, J. Rockstrom, and T. Oweis, (eds.) *Rainfed agriculture: unlocking the potential*. CAB International, Wallingford, Oxford, UK. pp. 276–291.

Katerji, N. & Rana, G. (2011). Crop reference evapotranspiration: A discussion of the concept, analysis of the process and validation. *Water Resources Management* 25, 6, 1581–1600.

Khouri, J., Amer, A. & Slih, A. (1995). *Rainfall water management in the Arab Region*. Prepared by UNESCO/ROSTA working group. United Nations Educational, Scientific and Cultural Organization/Regional Office for Science and Technology in Africa, Cairo, Egypt.

Klemm, W. (1990). *Bewässerung mit Niederschlagswasser ohne Zwischenspeicherung im Sahel. [Using runoff for irrigation without intermediate storage in the Sahel region]*. Doctoral thesis, Karlsruhe University, Karlsruhe, Germany. [In German].

Kumu, J.M. (1986). *Water harvesting for crop production in Baringo Pilot Semi-Arid Area Project (BPSAAP)*. Paper presented to the World Bank workshop on water harvesting in sub-Saharan Africa, Baringo, Kenya, 13–17 October 1986. World Bank, Washington D.C., USA.

Laing, I.A.F. (1981). Rainfall collection in Australia. In: G.R. Dutt, C.F. Hutchinson, and M. Anaya Garduno (eds.) *Rainfall collection for agriculture in arid and semi-arid regions: Proceedings of a workshop, University of Arizona, Tucson, Arizona 10–12 September 1980*. Commonwealth Agricultural Bureaux, Farnham House, Farnham Royal, Slough, UK. pp. 61–66.

Lélé, S.M. (1991). Sustainable development: A critical review. *World Development* 19, 607–621.

Li, F., Cook, S., Geballe, G.T. & Burch, W.R, Jr. (2000). Rainwater harvesting agriculture: An integrated system for water management on rainfed land in China's semiarid areas. *AMBIO* 29 (8), 477–483.

Linsley, R.K., Kohler, M.A. & Paulhus, J.L.H. (1982). *Hydrology for engineers (3rd ed.)*. New York: Mc Graw-Hill.

Malagnoux, M. (2009). Degraded arid land restoration for afforestation and agro-silvo-pastoral production through new water harvesting mechanized technology. In: *The future of drylands: International Conference on Desertification and Dryland Research, Tunis, Tunisia, 19–21 June 2006*. Springer. pp. 269–282.

Malesu, M.M., Oduor, A.R. & Odhiambu, O. (eds.) (2007). *Green water management handbook: Rainwater harvesting for agricultural production and ecological sustainability*. World Agroforestry Center (ICRAF) and Netherlands Ministry of Foreign Affairs, Nairobi, Kenya. Technical Manual 8.

Mati, B., De Bock, T., Malesu, M., Khaka, E., Oduor, A., Nyabenge, M. & Oduor, V. (2006). *Mapping the potentials for rainwater harvesting technologies in Africa: A GIS overview on development domains for the continent and ten selected countries*. World Agroforestry Centre (ICRAF) and Netherlands Ministry of Foreign Affairs, Nairobi, Kenya Technical Manual 6.

Meinzinger, F., Prinz, D. & Chahbani, B. (2004). *Analysis of traditional water harvesting techniques in south Tunisia. 4th World Water Congress: Innovation in drinking water treat-*

ment. 4th World Water Congress of the International Water Association (IWA), held in Marrakech, Morocco, 19–24 September 2004. Proceedings on CD-ROM. IWA Publishing, London, UK.

Morgan, P. (1990). *Rural water supplies and sanitation*. HongKong: Macmillan Education Limited, Hong Kong.

Murphy, J., Wallace, D. & Lane, L. (1977). Geomorphic parameters to predict hydrograph characteristics in the Southwest. *Water Resources Bulletin, American Water Resources Association,* 13 (1), 25–38.

Myers, L.E. & Frasier, G.W. (1975). Water harvesting — 2000 B.C. to 1974 A.D. In: Frasier, G.W. (ed.) *Proceedings, Water Harvesting Symposium, March 26–28, 1974, Phoenix, Arizona US Department of Agriculture, Agricultural Research Service, Berkeley, California, USA.* pp. 1–7.

Mzirai, O.B. & Tumbo, S.D. (2010). Macro-catchment rainwater harvesting systems: challenges and opportunities to access runoff. *Journal of Animal & Plant Sciences, 2010. Vol. 7, Issue 2*: 789–800. Access http://www.biosciences.elewa.org/JAPS

Nabhan, G. (1984). Soil fertility renewal and water harvesting in Sonoran desert agriculture: The Papago example. *Arid Lands Newsletter,* 20, 21–28.

Naudascher, E. (1996). Director, Institute for Hydromechanics, Karlsruhe University, Germany. (Personal communication June 1996).

Nasri, S. (2002). *Hydrological effects of water harvesting techniques*. Doctoral thesis. Department of Water Resources Engineering, Lund Institute of Technology, Lund, Sweden.

Nijhof, S. & Shrestha, B.R. (2010). Micro-credit and rainwater harvesting. In: *Proceedings of the IRC Symposium "Pumps, Pipes and Promises: Costs, Finances and Accountability for Sustainable WASH Services" 16–18 November 2010 in The Hague, NL.* [Online] Available from: http://www.rainfoundation.org/fileadmin/PublicSite/publications/IRC_symposium_nov.2010_Micro-credit_RainwaterHarvesting-Nepal.pdf [accessed 28th November 2010].

Nilsson, A. (1988). *Groundwater Dams for Small-Scale Water Supply*. London: Intermediate Technology Publ., London.

Nissen-Petersen, E. (2000). *Water from sand rivers. A manual on site survey, design, construction and maintenance of seven types of water structures in riverbeds*. RELMA Technical Handbook No. 23. Nairobi, Kenya: Sida Regional Land Management Unit.

Nissen-Petersen, E. (2006a). *Water from dry riverbeds*, Nairobi, Kenya: Danish International Development Assistance (Danida).

Nissen-Petersen, E. (2006b), *Water from rock outcrops. A handbook for engineers and technicians on site investigations, designs, construction and maintenance of rock catchment tanks and dams*. Nairobi, Kenya: ASAL Consultants Ltd. and Danish International Development Assistance (Danida).

Nissen-Petersen, E. (2006c). *Water from roads. A handbook for technicians and farmers on harvesting rainwater from roads*. Nairobi, Kenya: ASAL Consultants Ltd. and Danish International Development Assistance (Danida).

Nissen-Petersen, E. (2007). *Water from roofs. A handbook for technicians and builders on survey, design, construction and maintenance of roof catchments*. Nairobi, Kenya: ASAL Consultants Ltd. and Danish International Development Assistance (Danida).

NRCS (Natural Resources Conservation Services) Handbook (2008). *Hydrology*. USDA, Washington, D.C., USA.

Oberle, A. (2004). *GIS-based identification of suitable areas for various kinds of water harvesting in Syria*. Doctoral thesis. Karlsruhe University, Karlsruhe, Germany.

Oweis, T., Hachum, A. & Bruggeman, A. (eds.) (2004a). *Indigenous Water Harvesting Systems in West Asia and North Africa*. International Center for Agricultural Research in the Dry Areas, Aleppo, Syria.

Oweis, T., Hachum, A. & Bruggeman, A. (2004b). The role of indigenous knowledge in improving present water-harvesting practices., In: Oweis, T., Hachum, A. & Bruggeman, A. (eds.). *Indigenous Water-Harvesting Systems in West Asia and North Africa.* International Center for Agricultural Research in the Dry Areas, Aleppo, Syria. pp. 1–20

Oweis, T., Hachum, A. & Kijne, J. (1999). *Water harvesting and supplemental irrigation for improved water use efficiency in the dry areas.* SWIM Paper 7. International Water Management Institute, Colombo, Sri Lanka.

Oweis, T., Prinz, D. & Hachum, A. (2001). *Water harvesting: Indigenous knowledge for the future of the drier environments.* International Center for Agricultural Research in the Dry Areas, Aleppo, Syria.

Oweis, T. & Taimeh, A. (2001). Farm water-harvesting reservoirs: Issues of planning and management in dry areas. In: Z. Adeel (ed.). *Integrated land management in dry areas: Proceedings of a joint UNU-CAS international workshop, Beijing, China, 8–13 September 2001.* The United Nations University, Tokyo, Japan. pp. 165–182.

Pacey, A. & Cullis, A. (1986). *Rainwater harvesting.* The collection of rainfall and runoff in rural areas. Intermediate Technology Publications, London, UK.

Palmier, L.R. & Nobrega, R. (2010). The state-of-the-art in the cost-efficiency use of water harvesting techniques in agriculture. In: Steusloff, H. (ed.). *Integrated Water Resources Management: Proceedings of the International Conference IWRM Karlsruhe 2010.* Karlsruhe Institute of Technology (KIT) Scientific Publishing, Karlsruhe, Germany. pp. 65–70.

Pandey, D.N., Gupta, A.K. & Anderson, D.M. (2003). Rainwater harvesting as an adaptation to climate change. *Current Science*, 85 (1), 46–59.

Pandey, P.K., Soupir, M.L., Singh, V.P., Panda, S.N. & Pandey, V. (2011). Modeling rainwater storage in distributed reservoir systems in humid subtropical and tropical savannah regions. *Water Resources Management* 25 (13), 3091–3111.

Pachpute, J.S., Tumbo, S.D., Sally, H. & Mul, M.L. (2009). *Sustainability of rainwater harvesting systems in rural catchment of sub-Saharan Africa.* Water Resources Management, Vol. 23, No. 13, pp. 2815–2839.

Prinz, D. (1994). Water harvesting and sustainable agriculture in arid and semi-arid regions. In: C. Lacirignola & A. Hamdy (eds.). *Land and Water Resources Management in the Mediterranean Region, Vol. III: Proceedings of the CIHEAM Conference, 4–8 September, 1994, Valencano (Bari).* International Centre for Advanced Mediterranean Agronomic Studies (CIHEAM)/Mediterranean Agronomic Institute of Bari (IAM-B), Bari, Italy. pp. 745–762.

Prinz, D. (1996). Water harvesting: Past and future. In: L.S. Pereira, R.A. Feddes, J.R. Gilley, and B. Lesaffre (eds.). *Sustainability of irrigated agriculture.* NATO ASI Series, Series E: Applied Sciences 312, 137–144.

Prinz D. (2002a). *Multidisciplinary field programme review and development mission to western China. Report on soil and water conservation.* Food and Agriculture Organization of the United Nations, Bangkok.

Prinz, D. (2002b). The role of water harvesting in alleviating water scarcity in arid areas. Keynote Lecture, Proceedings, *International Conference on Water Resources Management in Arid Regions.* 23–27 March, 2002, Kuwait Institute for Scientific Research, Kuwait, (Vol. III, pp. 107–122).

Prinz, D. (2010). *Identification and design of potential water harvesting interventions in selected watersheds of northern Libya.* ARC Libya–ICARDA Collaborative Program, Water Harvesting and Irrigation Management Project, Tripoli, Libya. International Center for Agricultural Research in the Dry Areas, Aleppo, Syria.

Prinz, D. & Malik, A.H. (2002). More yield with less water—How efficient can be water conservation in agriculture? In: Tsakiris, G. (ed.) *Proceedings, 5th International. Conference on Water Resources Management in the Era of Transition, Athens, Greece, 4–8 September 2002.* European Water Resources Association, Athens, Greece. pp. 18–35.

Prinz, D., Wolfer, S., Tibebe, A. & Siegert, K. (1999). *Water harvesting for crop production.* [online] FAO Training Programme. Food and Agriculture Organization of the United Nations, Rome, Italy. Available from: http://www.fao.org/landandwater/aglw/wharv.htm. (Date of access 20th Feb. 2012).

Qiang, Z., Yuanhong, Li. & Chengxiang, M. (2007). *Rainwater harvesting*. Anhui Educational Publishing House, Hefei, Anhui, PR China.

Quilis, R.O., Hoogmoed, M., Ertsen, M., Foppen, J.W., Hut, R. & de Vries, A. (2009). Measuring and modeling hydrological processes of sandstorage dams on different spatial scales. *J. Phys. Chem. Earth.*, 34(4): 289–298. Available from: http://tudelft.academia.edu/MauritsErtsen/Papers/1461726/Accessed 15th Jan. 2012.

RAIN (Rainwater Harvesting Implementation Network) (2007). *A practical guide to sand dam implementation* [online]. RAIN Foundation, Amsterdam, NL. Available from: http://www.rainfoundation.org/fileadmin/PublicSite/Manuals/Sand_dam_manual_FINAL.pdf [date of access 6 September 2007].

RAIN (Rainwater Harvesting Implementation Network) (2008). *RAIN water quality guidelines.Guidelines and practical tools on rainwater quality.* [online] RAIN Foundation, Amsterdam, NL. Available from: www.rainfoundation.org/fileadmin/PublicSite/Manuals/RAIN_Rainwater_Quality_Policy_and_Guidelines_2009_v1.pdf [date of access 8 August 2009].

Raju, N.J., Reddy, T.V.K. & Munirathnam, P. (2006). Subsurface dams to harvest rainwater—a case study of the Swarnamukhi River basin, southern India. *Hydrogeology Journal*, 14(4), 526–531.

Rees, D., Qureshi, Z., Mehmood, S. & Raza, S. (1991). Catchment basin water harvesting as a means of improving the productivity of rain-fed land in upland Balochistan. *Journal of Agricultural Science* 116, 95–103.

Reij, C., Mulder, P. & Begemann, L. (1988). *Water harvesting for plant production.* World Bank, Washington, D.C., USA. World Bank Technical Paper 91.

Rocheleau, D., Weber, F. & Field-Juma, A. (1988). *Agroforestry in dryland Africa.* International Centre for Research in Agroforestry (ICRAF), Nairobi, Kenya.

Rockstroem, J. & Falkenmark, M. (2000). Semi-arid crop production from a hydrological perspective—Gap between potential and actual yields. *Critical Reviews in Plant Sciences* 19 (4), 319–346.

Rodriguez, A., Shah, N.A., Afazl, M., Mustafa, U. & Ali, I. (1996). Is water harvesting in valley floors a viable option for increasing cereal production in highland Balochistan, Pakistan? *Experimental Agriculture* 32 (3), 305–315.

Rwehumbiza, F.B., Mahoo, H.F. & Lazaro, E.A. (2000). Effective utilization of rainwater: More water from the same rain. In: Hatibu, N. & Mahoo, H.F. (eds.). *Rainwater harvesting for natural resources management. A planning guide for Tanzania.* RELMA Technical Handbook Series 22, Swedish International Development Cooperation Agency (Sida), RELMA, Nairobi, Kenya. pp. 23–35.

Sandford, S.G. (1983). *Management of pastoral development in the Third World.* John Wiley, London, UK.

Scanlon, B., Keese, K., Flint, A., Flint, L., Gaye, Ch., Edmunds, W. & Simmers, I. (2006). Global synthesis of groundwater recharge in semiarid and arid regions. *Hydrol. Process.* 20, 3335–3370.

Sengupta, N. (1993). *User-friendly irrigation designs.* London: Sage Publications, London, UK.

Shatta, A. & Attia, F. (1994). Environmental aspects of water harvesting. In: *Water harvesting for improved agricultural production: Proceedings of the FAO Expert Consultation, Cairo, Egypt, 21–25 November 1993.* Food and Agriculture Organization of the United Nations, Rome, Italy. pp. 257–270.

Siegert, K. (1994). Introduction to water harvesting. Some basic principles for planning. In: *Water harvesting for improved agricultural production: Proceedings of the FAO Expert Consultation, Cairo, Egypt, 21–25 November 1993*. Food and Agriculture Organization of the United Nations, Rome, Italy. pp. 9–21.

Somme, G., Oweis, T., Abdulal, A., Bruggeman, A. & Ali, A. (2004). *Micro-catchment water harvesting for improved vegetative cover in the Syrian badia*. International Center for Agricultural Research in the Dry Areas, Aleppo, Syria. On-farm Water Husbandry in WANA Series Report number: 3.

Steduto, P., Raes, D., Hsiao, T.C., Heng, L., Izzi, G. & Hoogeveen, J. (2008). *AquaCrop: A new model for crop prediction under water deficit conditions*. Options Méditerranéennes, Series A, No. 80, pp. 285–292.

Strangeways, I. (2006). *Precipitation: Theory, measurement and distribution*. Cambridge University Press, Cambridge, UK.

Tabor, J. & Djiby, B. (1987). *Soil and soil management for agriculture, forestry and range in Mauritania*. Mauritania Agricultural Research Project II. University of Arizona, Tucson, Arizona, USA.

Tauer, W. & Humborg, G. (1992). *Runoff irrigation in the Sahel zone: Remote sensing and geographic information systems for determining potential sites*. J. Margraf, Weikersheim, Germany.

Tauer, W. & Prinz, D. (1992). Runoff irrigation in the Sahel region: Appropriateness and essential framework conditions. In: Feyen, J., Mwendera, E. & Badji, M. (eds.) *Advances in planning, design and management of irrigation systems as related to sustainable land use: Proceedings of an International Conference, Leuven (Belgium), 14–17 September 1992*. pp. 945–953.

Tayaa, M. (1994). Present situation and prospects for improvement of water harvesting practices in Morocco. In: *Water harvesting for improved agricultural production. Proceedings of the FAO Expert Consultation, Cairo, Egypt, 21–25 November 1993*. Food and Agriculture Organization of the United Nations, Rome, Italy. pp. 231–254.

Thomas, T.H. & Martinson, D.B. (2007). *Roofwater harvesting: A handbook for practitioners*. Delft, The Netherlands, IRC International Water and Sanitation Centre. Technical Paper Series; no. 49.

UNEP (United Nations Environment Programme) (1983). *Rain and storm water harvesting in rural areas*. Tycooly International, Dublin, Ireland.

UNEP (United Nations Environment Programme) (2009). *Rainwater harvesting: a lifeline for human well-being. A report prepared by Stockholm Environment Institute*. [Online] United Nations Environment Program, Nairobi, Kenya. Available from: http://www.unwater.org/downloads/Rainwater_Harvesting_090310b.pdf [date of access 4th July 2009].

US Department of Agriculture (1992). *Agricultural waste management field handbook, Part 651*. National Engineering Handbook. Natural Resources Conservation Department, US Department of Agriculture, Washington, D.C., USA.

Van Dam, J.C. (2003). *Impacts of climate change and climate variability on hydrological regimes*. Cambridge University Press, Cambridge, UK.

Van Dijk, J.A. & Ahmed, M.H. (1993). *Opportunities for expanding water harvesting in sub-Saharan Africa: The case of teras of Kassala*. International Institute of Environment and Development, London, UK. Gatekeeper Series No. SA40.

Van Dijk, J. & Reij, Ch. (1994). Indigenous water harvesting techniques in sub-Saharan Africa: Examples from Sudan and the West African Sahel. In: *Water harvesting for improved agricultural production: Proceedings of the FAO Expert Consultation, Cairo, Egypt, 21–25 November 1993*. Food and Agriculture Organization of the United Nations, Rome, Italy. pp. 101–112.

Van der Zaag, P. & Gupta, J. (2008). Scale issues in the governance of water storage projects. *Water Resources Research* 44, W10417.

Van Steenbergen, F., Mehari Haile, A., Alemahayu, T., Alamirew, T. & Geleta, Y. (2011). Status and potential of spate irrigation in Ethiopia. *Water Resources Management* 25 (7), 1899–1913.

Vivian, G. (1974). *Conservation and diversion water control system in the Anansazi southwest.* Univ. of Arizona, Tuczon, Arizona, USA. Anthropological Papers Vol. 25.

Von Oppen, M. & Subba Rao, K.V. (1980). *Tank irrigation in semi-arid tropical India. Part I: Historical development and spatial distribution.* International Crops Research Institute for the Semi-Arid Tropics (ICRISAT), Patancheru, Andhra Pradesh, India. Economics Program Progress Report 5.

Walsh, J. (1991). *Preserving the options: Food productivity and sustainability.* Consultative Group on International Agricultural Research (CGIAR), CGIAR Secretariat, Washington, D.C.USA. Issues in Agriculture No. 2.

Werner, J. (1993). *Participatory development of agricultural innovations. Procedures and methods of on-farm research.* Schriftenreihe der Deutschen Gesellschaft für Technische Zusammenarbeit, Vol. 234. German Agency for Technical Cooperation (GTZ), Eschborn, Germany.

Western Australia, Central Regions Development Advisory Committee (1992). *Water harvesting: improving the reliability of farm dams in the Western Australian wheat belt.* Compiled by Brian Hillman. Dept. of Agriculture, report. 2.

WHO (World Health Organization) (2004). *Guidelines for drinking water quality (3rd ed.).* WHO, Geneva, Switzerland.

Worm, J. & van Hattum, T. (2006). *Rainwater harvesting for domestic use.* Agrodoc 43. Agromisa Foundation and CTA, Wageningen, The Netherlands.

Wright, P. (1985). *Soil and water conservation by farmers.* OXFAM, Ouagadougou, Burkina Faso.

Xiao-Yan, L., Zhong-Kui, X. & Xiang-Kui, Y. (2004). Runoff characteristics of artificial catchment materials for rainwater harvesting in the semiarid regions of China. *Agricultural Water Management* 65, 211–224.

Yahyaoui, H. & Ouessar, M. (2000). Abstraction and recharge impacts on the ground water in the arid regions of Tunisia: Case of Zeuss-Koutine water table. In: *Water management in dry zones: Proceedings of the International Workshop, Medenine, Tunisia, 18–22 October 1999.* UNU Desertification Series No. 2. The United Nations University, Tokyo, Japan. pp. 72–78.

Yuan, T., Fengmin, L. & Puhai, L. (2003). Economic analysis of rainwater harvesting and irrigation methods, with an example from China. *Agricultural Water Management* 60 (3–31), pp. 217–226

Yuanhong, L. & Qiang, Z. (2001). Mini-catchment technique for crop production and forestation in semiarid areas. In: *10th International Rainwater Catchment Systems Conference, "Rainwater International 2001", Mannheim, Germany, September 2001, Paper 4.9.* [Online] Available from: http://www.eng.warwick.ac.uk/ircsa/10th.html [Accessed 18th September 2001].

Zhu, K., Chen, H., Zhang, L. & Berndtsson, R. (1999). Disinfection and fresh-keeping of rainwater in small scale cisterns. In: R. Berndtsson (ed.). *Rain water harvesting and management of small reservoirs in arid and semi-arid areas. Proceedings of an international seminar,* 29.06-02.07.1998. Lund University, Report number: 3222, Lund, Sweden.

Index

Color plates

Figure 1.5 The *zay* (pitting) water harvesting system in Burkina Faso. The system concentrates runoff water in pits, where plants are grown. Photo courtesy E. Dudeck, GTZ.

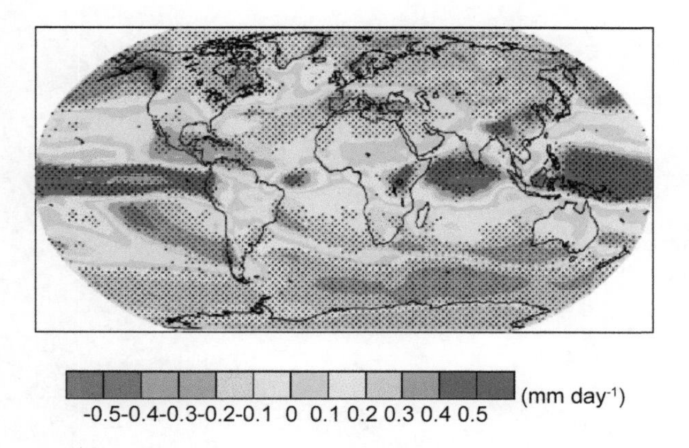

Figure 1.7 Predicted change in rainfall between the 20th and the 21st century (IPCC, 2007).

Figure 2.6 Discharge measuring of macro-catchments is a precondition for hydrological modeling. Measuring station in Oued Mina region in Algeria. Photo courtesy W. Klemm/Karlsruhe University, Germany.

Figure 2.7 Catchment area for a macro-catchment project in Mali, West Africa. It is extremely difficult, to calculate the runoff volumes from slopes like this one, covered with various types of vegetation, rock outcrops, etc. Additionally, the water yield varies over the rainy season with changing vegetation cover. Photo courtesy W. Klemm/Karlsruhe University, Germany.

Figure 3.1 Contour infiltration ditches combined with check dams (center), conserve soil and water and help recharge groundwater supplies. Photo courtesy A.K. Singh, Nirma University, India.

Figure 3.2 Channelling runoff water from long slopes (in the background) to fields, Kayes Province, Mali, West Africa. Photo courtesy W. Klemm/Karlsruhe University, Germany.

Figure 3.4 Various micro-catchment water harvesting systems applied at a research station: Contour bund water harvesting (right); inter-row (runoff strips) water harvesting for grain, pulse, and forage crops (center); and semicircular bunds for forage bushes (left). The uncultivated areas serve as catchments. Photo courtesy T. Oweis/ICARDA.

Figure B3.1.2 Drip irrigation is used to distribute the water within the greenhouse (Yuanhong & Qiang, 2001).

Figure 3.11 Inter-row (runoff strips) water harvesting on sloping land. At the right hand side semi-circular bunds. Photo courtesy T. Oweis/ICARDA.

Figure 3.12 Runoff-strip micro-catchment water harvesting showing small corrugation in the cropped area used to increase the uniformity of water distribution. Photo courtesy T. Oweis/ICARDA.

Figure 3.17 The *zay* pitting system in Burkina Faso, West Africa. The field is bunded to retain the runoff. Photo courtesy D. Prinz/Karlsruhe University, Germany.

Figure 3.18 Contour ridges. Photo courtesy T. Oweis/ICARDA.

Figure 3.19 Contour ridge water harvesting with fodder bushes. Photo courtesy T. Oweis/ICARDA.

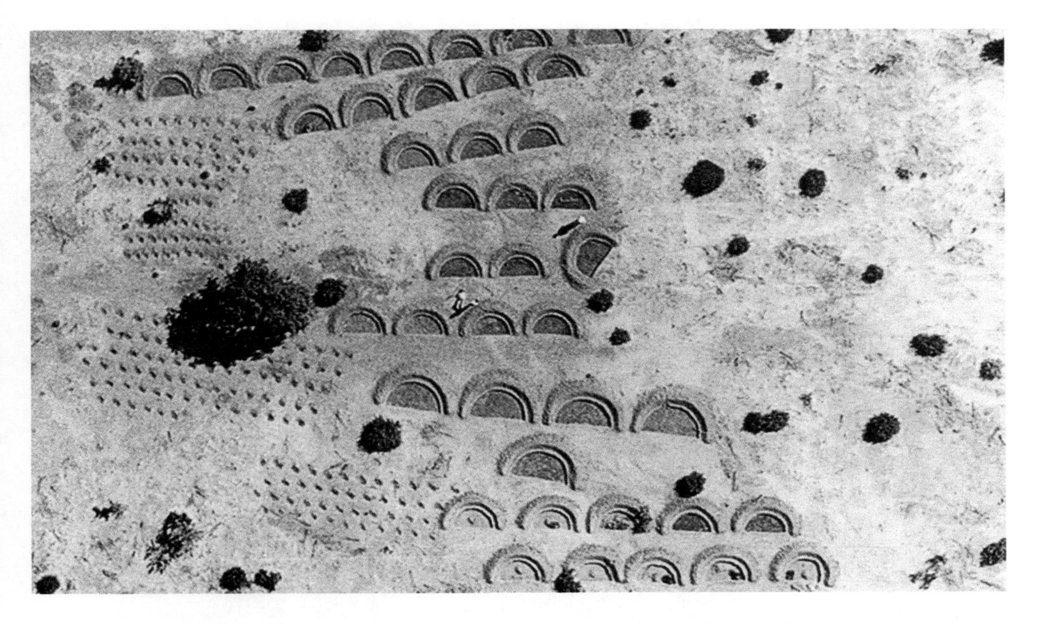

Figure 3.23 Micro-catchment water harvesting in Niger, including pits (left-hand side) and semicircular bunds. The larger semicircular bunds are 3 meters across (center); the smaller bunds are 2 meters across (lower two rows). Photo courtesy Deutsche Forschungsgemeinschaft (DFG).

Figure 3.24 Semicircular micro-catchments with fodder bushes in northwest Syria. Photo courtesy D. Prinz/Karlsruhe University, Germany.

Figure 3.25 Fig trees growing in eyebrow terraces near Salloum in northwestern Egypt. Photo courtesy T. Oweis/ICARDA.

(a)

(b)

(c)

Figure 3.27 Vallerani-type micro-catchments: (a) the 'Wavy Dolphin Plow' used to construct the micro-catchments; (b) The plow in use; basins are constructed along the contour; (c) micro-catchments after plant establishment. Photos courtesy by (a) D. Prinz/Karlsruhe University, Germany; (b) ICARDA (2006); (c) Photo courtesy T. Oweis/ICARDA.

(a) (b)

Figure 3.28 The Vallerani-type bunds look like an intermittent contour ridge. Mechanization facilitates construction of these bunds on a large scale. Photos courtesy T. Oweis/ICARDA.

Figure 3.33 Hillside conduit system in Mali: Flooded fields with a diversion. (Klemm, 1990).

Figure 3.37 Typical *tabia* in southern Tunisia. Photo courtesy T. Oweis/ICARDA.

Figure 3.39 A *wadi*-bed water harvesting system in Tunisia. Photo courtesy T. Oweis/ICARDA.

Figure 3.40 Stone walls constructed using gabions across a *wadi*-bed in Marsa Matruh area in northwest Egypt. Photo courtesy T. Oweis/ICARDA.

Figure 3.42 An example of warping from the loess plateau in central China. (Yuanhong & Qiang, 2001).

Figure 4.1 Using a roller to compact runoff strips between crop strips has high capital cost but low labor requirement. Photo courtesy T. Oweis/ICARDA.

Figure 4.4 A roaded catchment with a reservoir in Australia (Western Australia, Central Regions Development Advisory Committee, 1992).

Figure B5.1.1 Topographic features, infrastructure, and even vegetation show up in high-quality images from Google Earth. (Prinz, 2010).

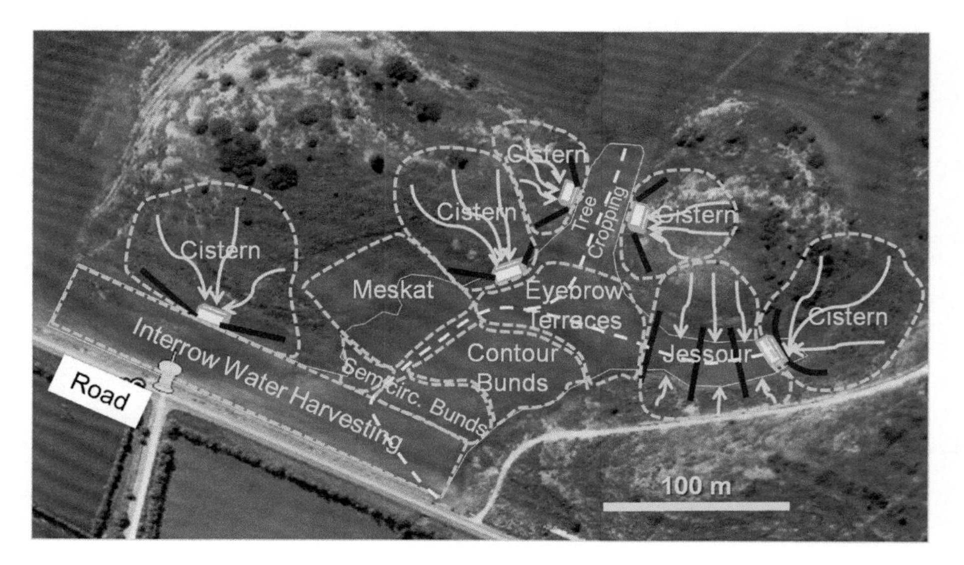

Figure B5.1.2 A Google Earth image used to plan water harvesting interventions in Libya. (Prinz, 2010).

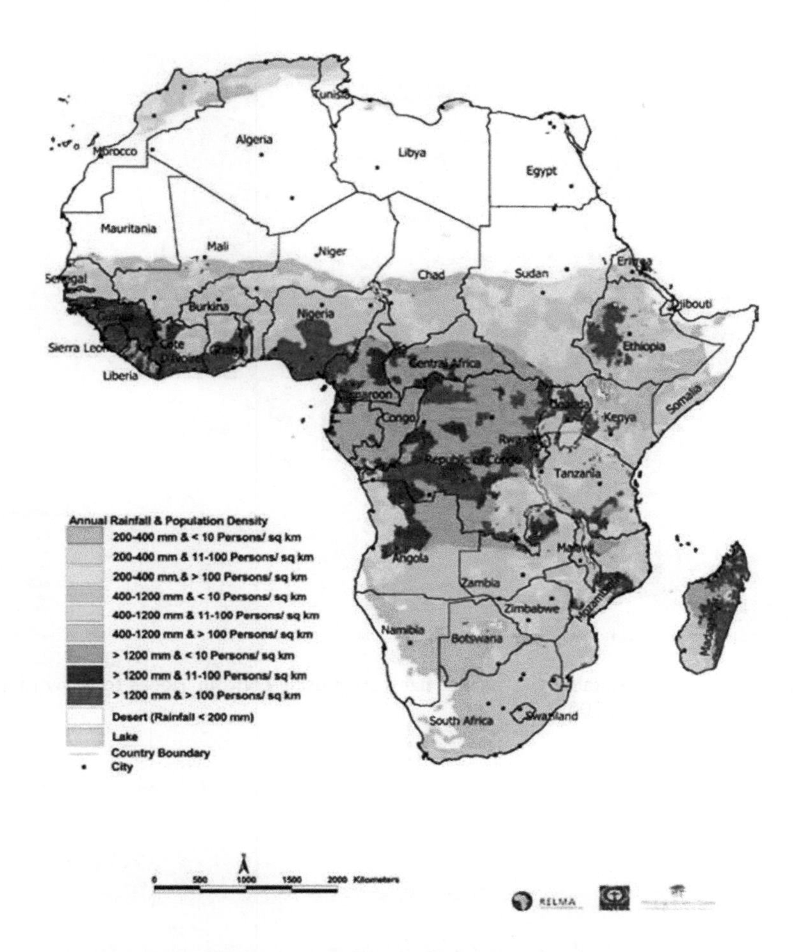

Figure B5.2.1 Map showing potential for runoff-water harvesting across Africa. (Mati *et al.*, 2006).

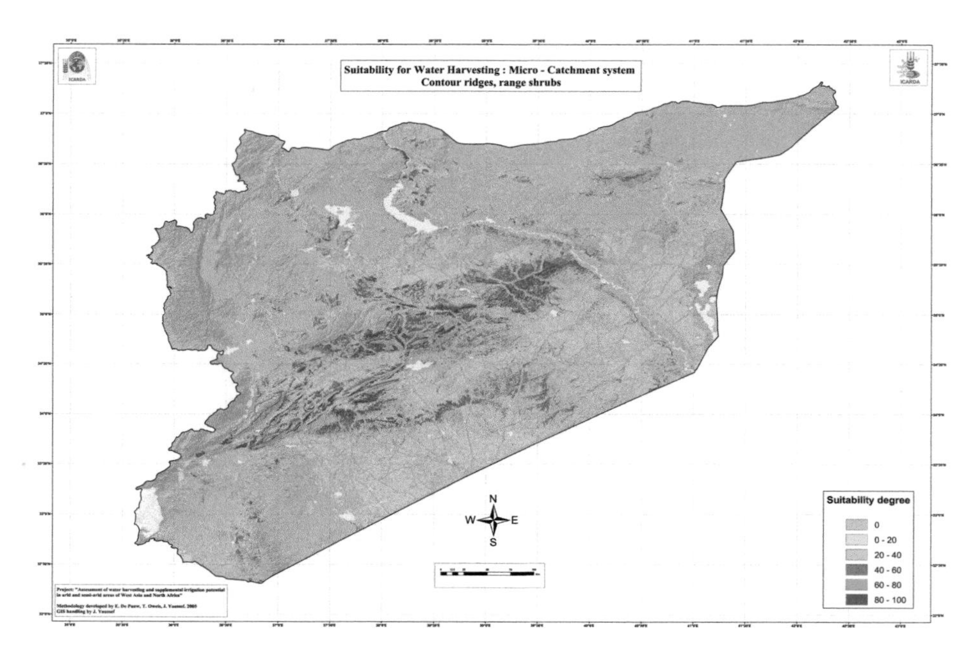

Figure B5.3.1 Map showing suitability for water harvesting using contour ridges to grow range shrubs (De Pauw *et al.*, 2009).

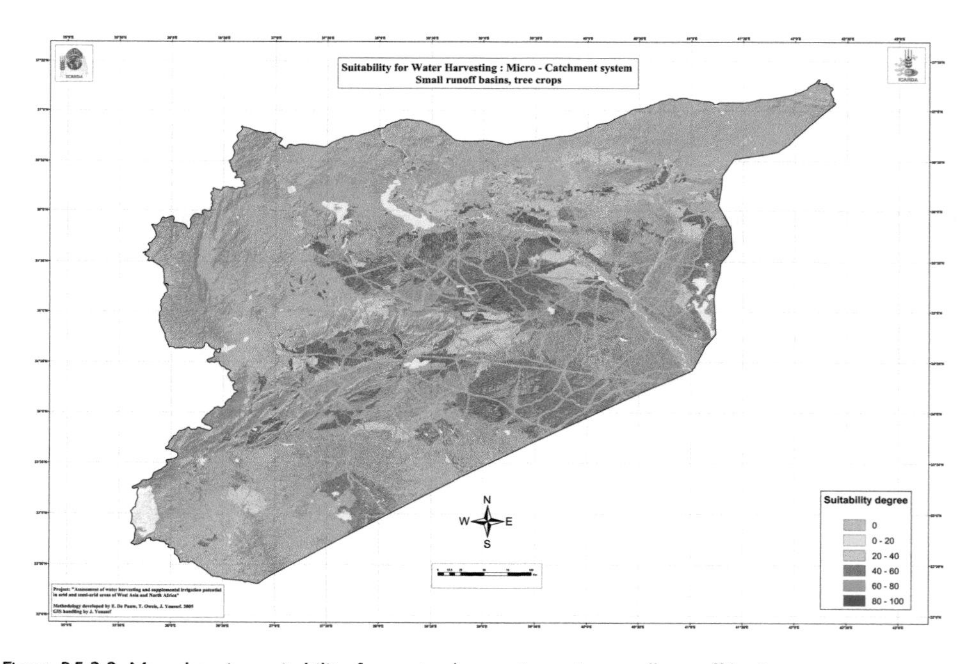

Figure B5.3.2 Map showing suitability for water harvesting using small runoff basins to grow tree crops (De Pauw *et al.*, 2009).

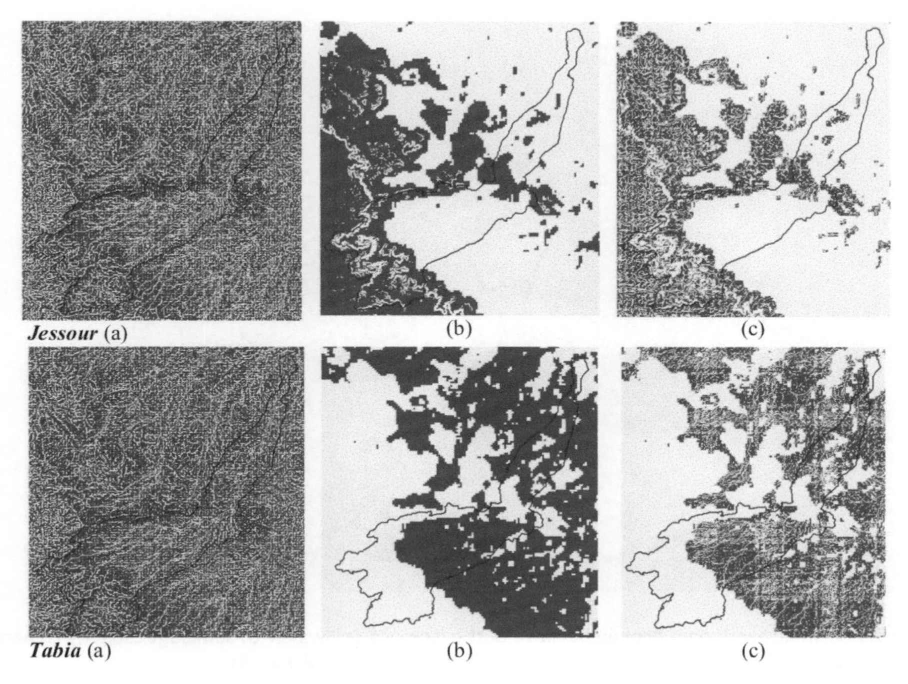

Figure B5.4.1 Potential for *jessour* (a) and *tabias* (b) in the study area (green color) according to flow accumulation and slopes. (1) Using only water flow accumulation, (2) Using only slope, (3) Using both flow accumulation and slope. (De Pauw *et al.*, 2009).

Figure 7.8 Small farm reservoir for supplemental irrigation in Tunisia. Photo courtesy S. Wolfer/Karlsruhe University, Germany.

Figure 7.9 A small earth dam broken because of the lack of a spillway with sufficient capacity to handle peak overflow. Photo courtesy State Dept. of Western Australia.

Figure 7.10 A masonry dam used to harvest floodwater in a *wadi* in northwestern Egypt. The water is used for supplemental irrigation of field crops and groundwater recharge. Photo courtesy T. Oweis/ICARDA.

Figure 7.12 A *hafair* in Jordan. The water in the *hafair* will be available for 3 months after the end of the rainy season. Photo courtesy T. Oweis/ICARDA.

Figure B7.2.1 Sedimentation basin and cistern in Gansu Province, PRI China. (Yuanhong & Qiang, 2001).

Figure B7.2.2 'Water cellar' (foreground) in Gansu Province, PR China. (Yuanhong & Qiang, 2001).

Figure B7.3.2 Water passes through a silt trap before entering the cistern in Gansu Province, PR China. (Yuanhong & Qiang, 2001).

Figure 8.7 A measuring weir in a *wadi* used to monitor runoff in the Syrian steppe. Photo courtesy T. Oweis/ICARDA.

Figure 8.8 Measuring the runoff in a small channel in the Syrian steppe. The small tank in the ground can store about 1 m^3 of runoff. It is advisable to plaster the soil surface around the tank to avoid soil erosion and mistakes in runoff determination. Guiding dikes may be provided on both sides upstream the tank to force all flow to pass over the tank and to prevent erosion on both sides of the tank. Photo courtesy T. Oweis/ICARDA.

Figure 9.1 Recharging groundwater by harvesting floodwater in Saudi Arabia. An assessment of the economic benefits of such a project is hardly feasible. Photo courtesy A.-M. Al Sheikh, Saudi Arabia.

Figure 10.1 Cistern in NE Libya. The water from the cistern is pumped into the trough (foreground) to supply goats, sheep, and other animals. To prevent contamination of the cistern water, the trough should be located several meters apart. Photo courtesy D. Prinz/Karlsruhe University, Germany.

Figure 10.2 A flooded *wadi* in southern Morocco. Each flooding is associated with sediment transport increasing the turbidity of the water. Photo courtesy D. Prinz/Karlsruhe University, Germany.

Figure 10.3 Rainwater tank in Amhara region, Ethiopia. The water collected from the rooftop is used mainly for domestic purposes, sometimes for gardening. Photo courtesy E.G. Zerihun/ Water Resources Development Program, Organization for Rehabilitation and Development in Amhara (ORDA), Bahir Dar, Ethiopia.

Figure 10.4 A constructed catchment with cistern to cover the drinking and domestic water demand of a small Tunisian village. Photo courtesy D. Prinz/Karlsruhe University, Germany.

Figure 10.5 In this example from Ethiopia, runoff water has to pass through two settling basins before flowing into the sealed pond. Photo courtesy G. Zerihun/ORDA, Bahir Dar, Ethiopia.

Figure 10.6 Flow of water after a rainfall event to an ephemeral river bed in southern Tunisia. The water moves soil particles, which may carry nutrients as well as chemical pollutants. Photo courtesy B. Chahbani/IRA, Medenine, Tunisia.

Figure 10.7 Runoff water flowing through a culvert after a rainfall event in the Syrian steppe. The water is rich in sediments, which settle further downstream. Photo courtesy T. Oweis/ICARDA.

Figure 10.8 The sediments carried by the flowing water will settle and change the morphology of the ephemeral river course. Photo courtesy D. Prinz/Karlsruhe University, Germany.

For Product Safety Concerns and Information please contact our EU representative GPSR@taylorandfrancis.com Taylor & Francis Verlag GmbH, Kaufingerstraße 24, 80331 München, Germany

T - #0179 - 160425 - C26 - 246/174/14 [16] - CB - 9780415621144 - Gloss Lamination